The Air Force Health Study
ASSETS RESEARCH PROGRAM

I0896215

Committee on the Management of the Air Force Health Study
Data and Specimens—Report to Congress

Board on the Health of Select Populations

INSTITUTE OF MEDICINE
OF THE NATIONAL ACADEMIES

THE NATIONAL ACADEMIES PRESS
Washington, D.C.
www.nap.edu

THE NATIONAL ACADEMIES PRESS 500 Fifth Street, NW Washington, DC 20001

NOTICE: The project that is the subject of this report was approved by the Governing Board of the National Research Council, whose members are drawn from the councils of the National Academy of Sciences, the National Academy of Engineering, and the Institute of Medicine. The members of the committee responsible for the report were chosen for their special competencies and with regard for appropriate balance.

This study was supported by Grant No. HT9404-12-1-0028 between the National Academy of Sciences and the Department of Veterans Affairs. Any opinions, findings, conclusions, or recommendations expressed in this publication are those of the author(s) and do not necessarily reflect the view of the organizations or agencies that provided support for this project.

International Standard Book Number-13: 978-0-309-33914-8
International Standard Book Number-10: 0-309-33914-6

Additional copies of this report are available for sale from the National Academies Press, 500 Fifth Street, NW, Keck 360, Washington, DC 20001; (800) 624-6242 or (202) 334-3313; http://www.nap.edu.

For more information about the Institute of Medicine, visit the IOM home page at: **www.iom.edu.**

Printed in the United States of America

The serpent has been a symbol of long life, healing, and knowledge among almost all cultures and religions since the beginning of recorded history. The serpent adopted as a logotype by the Institute of Medicine is a relief carving from ancient Greece, now held by the Staatliche Museen in Berlin.

Suggested citation: IOM (Institute of Medicine). 2015. *The Air Force Health Study Assets Research Program.* Washington, DC: The National Academies Press.

"Knowing is not enough; we must apply.
Willing is not enough; we must do."
—Goethe

INSTITUTE OF MEDICINE
OF THE NATIONAL ACADEMIES

Advising the Nation. Improving Health.

THE NATIONAL ACADEMIES

Advisers to the Nation on Science, Engineering, and Medicine

The **National Academy of Sciences** is a private, nonprofit, self-perpetuating society of distinguished scholars engaged in scientific and engineering research, dedicated to the furtherance of science and technology and to their use for the general welfare. Upon the authority of the charter granted to it by the Congress in 1863, the Academy has a mandate that requires it to advise the federal government on scientific and technical matters. Dr. Ralph J. Cicerone is president of the National Academy of Sciences.

The **National Academy of Engineering** was established in 1964, under the charter of the National Academy of Sciences, as a parallel organization of outstanding engineers. It is autonomous in its administration and in the selection of its members, sharing with the National Academy of Sciences the responsibility for advising the federal government. The National Academy of Engineering also sponsors engineering programs aimed at meeting national needs, encourages education and research, and recognizes the superior achievements of engineers. Dr. C. D. Mote, Jr., is president of the National Academy of Engineering.

The **Institute of Medicine** was established in 1970 by the National Academy of Sciences to secure the services of eminent members of appropriate professions in the examination of policy matters pertaining to the health of the public. The Institute acts under the responsibility given to the National Academy of Sciences by its congressional charter to be an adviser to the federal government and, upon its own initiative, to identify issues of medical care, research, and education. Dr. Victor J. Dzau is president of the Institute of Medicine.

The **National Research Council** was organized by the National Academy of Sciences in 1916 to associate the broad community of science and technology with the Academy's purposes of furthering knowledge and advising the federal government. Functioning in accordance with general policies determined by the Academy, the Council has become the principal operating agency of both the National Academy of Sciences and the National Academy of Engineering in providing services to the government, the public, and the scientific and engineering communities. The Council is administered jointly by both Academies and the Institute of Medicine. Dr. Ralph J. Cicerone and Dr. C. D. Mote, Jr., are chair and vice chair, respectively, of the National Research Council.

www.national-academies.org

COMMITTEE ON THE MANAGEMENT OF THE AIR FORCE HEALTH STUDY DATA AND SPECIMENS—REPORT TO CONGRESS

DAVID J. TOLLERUD (*Chair*), Professor and Chair, Department of Environmental and Occupational Health Sciences, University of Louisville
MARY N. HAAN, Professor and Vice Chair, Department of Epidemiology and Biostatistics, University of California, San Francisco
KENNETH W. KIZER, Distinguished Professor, University of California, Davis, School of Medicine, and Betty Irene Moore School of Nursing; Director, Institute for Population Health Improvement, UC Davis Health System
DIANE LEONG, Director, Biobanking and Human Biological Sample Management, Research Biology Department, Gilead Sciences, Inc.
BRADLEY MALIN, Associate Professor and Vice Chair, Biomedical Informatics, Vanderbilt University
LISA M. McSHANE, Mathematical Statistician, Biometric Research Branch, National Cancer Institute
F. JAVIER NIETO, Professor and Chair, Department of Population Health Sciences; Helfaer Professor of Public Health, School of Medicine and Public Health, University of Wisconsin–Madison
MICHAEL A. STOTO, Professor of Health Systems Administration and Population Health, Georgetown University
LINDA D. YOUNGMAN, Branch Chief, Community Grants and Program Development Center for Substance Abuse Prevention, Substance Abuse and Mental Health Services Administration

Study Staff

ANNE STYKA, Study Director
DAVID A. BUTLER, Scholar
HARRIET CRAWFORD, Data Operations Manager
VICTORIA KING, Senior Programmer/Analyst
DWAYNE BELL, Programmer/Analyst
SULVIA DOJA, Senior Program Assistant
JULIE WILTSHIRE, Financial Associate
FREDERICK ERDTMANN, Director, Board on the Health of Select Populations

Reviewers

This report has been reviewed in draft form by persons chosen for their diverse perspectives and technical expertise, in accordance with procedures approved by the National Research Council's Report Review Committee. The purposes of this independent review are to provide candid and critical comments that will assist the institution in making its published report as sound as possible and to ensure that the report meets institutional standards for objectivity, evidence, and responsiveness to the study charge. The review comments and draft manuscript remain confidential to protect the integrity of the deliberative process. We wish to thank the following individuals for their review of this report:

Garnet L. Anderson, Fred Hutchinson Cancer Research Center
Linda M. Bartoshuk, University of Florida
Carolyn C. Compton, Arizona State University
Helena J. Ellis, Duke University
Kelly Edwards, University of Washington
David A. Kalman, University of Washington
James W. McNally, University of Michigan
Michael D. Parkinson, University of Pittsburgh Medical Center

Although the reviewers listed above have provided many constructive comments and suggestions, they were not asked to endorse the conclusions or recommendations, nor did they see the final draft of the report before its release. The review of the report was overseen by **Dan G. Blazer,** Duke University Medical Center, and **Mark R. Cullen,** Stanford University. Appointed by the National

Research Council and the Institute of Medicine, they were responsible for making certain that an independent examination of the report was carried out in accordance with institutional procedures and that all review comments were carefully considered. Responsibility for the final content of the report rests entirely with the authoring committee and the institution.

Preface

Well designed, longitudinal prospective epidemiologic studies that follow a population for many years and that include the collection and storage of biospecimens for use in current and future research are relatively rare. Such epidemiologic studies, such as the Framingham Heart Study, have gained fame because they are relatively few in number and are associated with vast amounts of data useful for many types of research. Another member of this select group is the Air Force Health Study (AFHS), which is not well recognized outside of the veterans health research community. The extensive data and associated biospecimens collected during its 20-year course have been preserved and continue to be valuable for many types of research and many scientific disciplines.

The 2007 National Defense Authorization Act directed the Air Force to transfer custody of the AFHS assets to the Medical Follow-Up Agency of the Institute of Medicine (IOM). With the passage of the Veterans' Benefits Improvement Act in 2008, the IOM was directed to undertake a program of research using the data and biospecimens to better understand the determinants of health and promote wellness in veterans and the general population. Because the AFHS has only been accessible to the scientific community for research for less than 3 years, the vast majority of its potential is yet to be realized.

The committee gratefully acknowledges the many individuals and groups that generously shared their time, expertise, and insights with the committee. Among these numerous people, the committee would like to acknowledge the veterans who continue to allow their materials to be used for research, the investigators who have demonstrated the value and some of the many uses of the AFHS assets in cutting-edge research, the potential researchers, the Department of Veterans

Affairs who sponsored this program and continues to publicize its availability, persons from the many national professional organizations who have promoted the availability of the AFHS assets for research, and the Air Force Research Laboratory staff who have maintained the biospecimens. The committee is also grateful for the service and insight provided by Dr. Marie A. Bernard, Deputy Director of the National Institute on Aging of the National Institutes of Health, and Dr. Richard J. Jackson, Professor and Chair of Environmental Health Sciences at the Fielding School of Public Health at the University of California, Los Angeles, who served on the advisory committee but rotated off before the initiation of the consensus committee.

I would especially like to acknowledge the tremendous work of the IOM staff who have worked tirelessly with the veterans, researchers, and committee to make the AFHS Assets Research Program a successful pilot research program. The committee is grateful to Anne Styka, who served as study director for this project, and to the IOM staff members who contributed to the project: David Butler, Harriet Crawford, Victoria King, Dwayne Bell, and Sulvia Doja.

David J. Tollerud, *Chair*
Committee on the Management of the Air Force Health Study
Data and Specimens—Report to Congress

Contents

APPENDIXES

Abbreviations and Acronyms

AFHS	Air Force Health Study
AFRL	Air Force Research Laboratory
CDC	Centers for Disease Control and Prevention
CI	confidence interval
CRADA	cooperative research and development agreement
ECG	electrocardiogram
GAO	Government Accountability Office
ICD-9	*International Classification of Diseases, Ninth Revision*
ICPSR	Inter-university Consortium for Political and Social Research
IOM	Institute of Medicine
IRB	institutional review board
MFUA	Medical Follow-Up Agency
MGUS	monoclonal gammopathy of undetermined significance
NCI	National Cancer Institute
NHLBI	National Heart, Lung, and Blood Institute
NIH	National Institutes of Health

OSTP	Office of Science and Technology Policy
RFP	request for proposals
RR	relative risk
SAIC	Science Applications International Corporation
SAS	Statistical Analysis System
TCDD	2,3,7,8-tetrachlorodibenzo-*p*-dioxin
VA	Department of Veterans Affairs
WPAFB	Wright-Patterson Air Force Base

Summary

Longitudinal prospective studies that follow a population for a decade or more are the gold standard of observational epidemiologic studies. Such studies are uncommon, though, because of the great expense and time needed to conduct them. Rarer still are well-designed cohort studies that include the collection and storage of biospecimens for use in current and future analyses. The Framingham Heart Study and Atherosclerosis Risk in Communities Study are two examples of such research efforts that were designed to examine a particular health outcome but used to produce broader research findings as well. The design, data and biospecimen collection, and continued use of the assets assembled in these studies are considered to be exemplars of epidemiologic research.

One member of this select group—little known outside the veterans health research community—is the Air Force Health Study (AFHS): a longitudinal, prospective epidemiologic study of more than 2,700 men followed for approximately 20 years. This cohort participated in up to six intensive physical examinations with high rates of compliance. In addition to a complete record of clinical measurements and observations collected at these exams, serum and other biological samples were obtained and preserved. Extensive questionnaires addressing health, lifestyle, and socioeconomic status were administered during each exam, and other information was obtained about the participants' employment, families and offspring, and potential sources of environmental exposures.

The U.S. Congress, which played a role in the initiation of the study in 1979, passed three laws with sections relating to the disposition of its assets (data and biospecimens). Section 602 of Public Law 108-183 directed the Secretary of Veterans Affairs to contract with the National Academy of Sciences' Institute of Medicine (IOM) to address several questions regarding the appropriate management of the AFHS assets at the conclusion of the study. The committee formed to answer them concluded that the AFHS data and biospecimens appeared to be of high quality and well preserved. It recommended that these assets be maintained and used to foster new research that addressed a broader range of questions than the original study and that took advantage of advances in science and technology to learn more from the collected assets. However, it also noted that obstacles existed to making the assets available to outside investigators; in particular, deficiencies in their documentation and organization. The committee identified several possible appropriate custodians for the data and biospecimens, and recommended that a mechanism of independent oversight on all future research using the AFHS assets be implemented to ensure that the work conducted on them was of high scientific quality and conformed to legal requirements and ethical standards for the handling of the participants' information (IOM, 2006).

Following on the findings, conclusions, and recommendations from the 2006 IOM report, Congress passed Public Law 109-364, Section 714 of which directed the U.S. Air Force to transfer custodianship of the AFHS's data and biospecimens to the Medical Follow-Up Agency (MFUA)—a unit within the IOM responsible for data-intensive epidemiologic studies—at the completion of the study in September 2007. Before the assets could be transferred to MFUA, the surviving members of the cohort were asked to consent to the transfer of their data and biospecimens to the custody of the IOM and more than 90% gave this permission.

Two years after MFUA became the custodian of the AFHS assets, Congress passed Public Law 110-389, Section 803 which directed the Department of Veterans Affairs (VA) to provide funding to MFUA to maintain and manage the AFHS assets and to make them "available as a resource for future research for the benefit of veterans and their families, and for other humanitarian purposes." MFUA was authorized to administer a research program aimed at "understanding the determinants of health, and promoting wellness," specifically in veterans but also for the broader U.S. population. The law also directed that up to $250,000 per year be used for pilot funds to be made available when appropriate to support investigations using the AFHS assets. At the end of that effort, Congress directed that a report be submitted to them that considered the feasibility and advisability of conducting additional research on the AFHS data and biospecimens. This report responds to that congressional directive.

THE COMMITTEES' TASKS

The IOM undertook two related efforts to carry out the tasks identified in Public Law 110-389. First, in its role as custodian of the AFHS data and biospecimens, it established an advisory committee to facilitate new research using the AFHS assets and to review proposals from researchers who requested access to the data and biospecimens. Second, to conform to institutional requirements, a consensus committee composed of the same membership was created to write the report regarding the feasibility and advisability of conducting additional research on the AFHS data and biospecimens called for in the legislation.

The advisory committee operationalized its responsibilities by developing and implementing protocols for

- reviewing the content of prospective research proposals;
- determining whether a proposal met the criteria for scientific merit;
- overseeing the dissemination and use of the biospecimens, including deciding whether and which specimens should be made available and how much material should be provided; and
- awarding pilot funding to investigators who applied for support.

The statement of task for the consensus committee—the body responsible for this report—is shown in Box S-1. Its work comprised a summary of the activities carried out since the IOM assumed custodianship of the AFHS assets, a critical analysis of the lessons learned in administering these assets, and a survey of the experiences of the investigators who received data or specimens to solicit feedback on the process. The committee also used the findings and recommendations of the *Disposition of the Air Force Health Study* report as a guide for assessing what was believed to be feasible and possible for a research program managing and using the AFHS assets and its realized experience.

BOX S-1
Statement of Task—Consensus Committee

A June 28, 2011, contract between the National Academies and the Department of Veterans Affairs entitled *Air Force Health Study (AFHS) Assets Research Program*—which operationalizes directives contained in Public Law 110-389, section 803 (Maintenance, Management, and Availability for Research of Assets of Air Force Health Study)—directs the IOM to "provide the final report to Congress at the end of the final period of performance outlining the feasibility and advisability of maintaining the biospecimens based on interest generated from the general scientific community and results of pilot projects and other research projects using the AFHS assets." An ad hoc committee will conduct a study and prepare a report in response to this charge. The final period of performance ends on June 30, 2015.

THE AIR FORCE HEALTH STUDY

The original purpose of the AFHS was to determine whether Air Force personnel who had participated in the Operation Ranch Hand program of wartime herbicide spraying in Vietnam had experienced adverse health outcomes as a result of their service. Personnel who were involved in the operation—commonly referred to as the Ranch Hands—and who were living when the study began were identified and individually matched on age, race, and military occupation to comparison subjects who were stationed in and flew cargo operations elsewhere in Southeast Asia during the Vietnam conflict, had not been exposed to tactical herbicides, and were assumed to be similar in training profiles, socioeconomic factors, and lifestyles.

The AFHS program was conducted over 27 years (1979–2006). Clinical measurements and observations were made, extensive health questionnaires were administered, and biological samples were obtained from the participants at each of six exams, or *cycles*: at baseline (1982) and then in years 3, 5, 10, 15, and 20 of the study. Data collected during physical examinations comprised indices of health status that encompassed general health and endpoints by major organ system. More than 200 clinical laboratory tests and measures were conducted and evaluated, and more than 60 of these were measured at all six cycles. Other information was obtained from personal medical records and from the participants' families. Box S-2 summarizes the major types of information collected.

BOX S-2
Summary of Clinical, Laboratory, and Questionnaire Information Collected by the Air Force Health Study

Biospecimens: serum, whole blood, adipose tissue, semen, urine

Clinical measures: spirometry, chest radiographs, electrocardiograms, psychological testing, dermatology exams, peripheral vascular exams, neurological assessments

Demographics: education, employment, income, military experience, age, race

Endpoints by organ system: neurologic, endocrine, pulmonary, immunologic, renal, gastrointestinal, hepatic, hematologic, dermatologic, psychiatric, neoplasia, cardiovascular, reproductive

Familial factors: marital and fertility history, child and family health

General health: health habits, physical activities, leisure activities, toxic substances exposures, vital status, cause of death

While the study was completed in 2006, the extensive health data linked to several types of longitudinally collected biologic specimens—some 91,000 serum, whole blood, urine, semen, and adipose tissue specimens—remain a resource for additional research. The AFHS assets are exceptional in the sheer multitude and range of types of information available for each participant.

TRANSFER, PREPARATION, AND MANAGEMENT OF THE AFHS ASSETS

Following passage of Public Laws 109-364 and 110-389, it took approximately 3 years for a final agreement to be reached and for funding to be in place for administration of the new AFHS Assets Research Program through the IOM. More than 123,000 electronic files were transferred to MFUA. These comprise text files, images of X-rays and electrocardiograms, analysis and translation files (Statistical Analysis System databases and programs, Fortran programs, Excel spreadsheets), and PDFs of questionnaires, physical exam forms, reports, and other paper records. Because the IOM does not have the physical plant needed to maintain a biospecimens collection, MFUA reached an agreement and contracted with the Air Force Research Laboratory (AFRL) at Wright-Patterson Air Force Base in Dayton, Ohio, to house the biospecimens.

Although 2,758 individuals participated in at least one exam cycle, not all gave consent to have their data and biospecimens transferred to the IOM or to be used in future research. Materials from the 2,488 participants who allowed the custody change were transferred. Subsequently, the IOM sought permission from the participants to allow use of their materials in future research, and 2,210 gave their consent.

The electronic data transferred to the IOM were not standardized between cycles or in a database format. The first year of MFUA's management of the data was, therefore, dedicated to documenting, reading in, verifying, and integrating the disparate files, and otherwise preparing them for use by outside investigators as well as obtaining the consent of the cohort. Further, an updated database of the biospecimens needed to be created to document accessions of them. This was a unique and challenging undertaking. To the committee's knowledge, such a large transfer of both data and biospecimens and the need to recreate such an extensive database with little documentation and no formal consultants from the original project staff had never been accomplished before. Because of this, there were no examples for the IOM staff or the advisory committee to base such work on and not all issues could be anticipated when the program first began. Questions regarding the actual quality and usability of the data and biospecimens needed to be addressed before research using the AFHS assets could be realized.

While these activities were under way, the advisory committee charged with administrative oversight of the research program and fostering the use of the AFHS data and biospecimens in new research established a process for soliciting and reviewing research proposals. The first request for proposals (RFP) was announced on April 30, 2012. International and U.S. researchers from academic institutions, industry, and government organizations were eligible to apply for access to the AFHS assets. The application process consisted of two steps: a letter of intent and a full proposal. Proposals were required to be hypothesis-driven studies, not merely data mining endeavors. The RFP was publicized on the AFHS project page of the IOM website and initially sent to researchers and organizations that were familiar with the AFHS or had an interest in research related to herbicide exposure.

Since MFUA became the custodian of the AFHS assets, more than 80 inquiries from researchers have been received. As of February 1, 2015, 26 letters of intent were submitted by investigators, 24 of whom were invited to submit full proposals. Of the 24 proposal invitations, 19 complete proposals were submitted to the committee for review. Of the 19 proposals reviewed by the committee, 13 proposals were approved; 6 studies required access to the data only, and 7 studies used both data and biospecimens.

NEW RESEARCH USING THE AFHS ASSETS

The first proposals were approved in 2012, and while it is premature to expect publications so soon into the program, the committee was able to review preliminary results from approved studies in the feasibility stage, and one investigator published two papers describing his results. The research proposals reviewed and approved for use of the AFHS biospecimens and data span a wide range of scientific and biomedical research topics (see Table S-1).

Awareness of the AFHS resource and its potential value beyond herbicide-exposure research has been disseminated throughout the research community, as evidenced by a steady increase in the diversity and number of proposals received. Research topics that have emerged among the submitted and approved proposals include aging, neurologic diseases, and cancer research, demonstrating that AFHS data and biospecimens remain relevant for important biomedical and public health research projects. Moreover, the biospecimens have been well-preserved and viable, and given the current storage conditions, should remain so for many years to come. The extensive data and biospecimens collected over the extended period of the AFHS makes it invaluable for examining potential etiological factors relating to rare diseases, diseases and conditions that have a long latency periods, or for examining biomarkers associated with the natural history of slowly progressive disorders. For example, diagnostic data can be combined with DNA analysis from whole blood, semen, or adipose tissue samples to investigate genetic markers of disease predisposition or prognosis.

The committee solicited comments from the principal investigators of approved research studies after they had received all of their requested data or biospecimen samples. Investigators were asked to give specific examples of how

TABLE S-1 Studies Approved for Use of the AFHS Data or Biospecimens Since 2012[a,b]

Principal Investigator	Institution	Proposal Title	Request Type
Batty	University of Edinburgh	Cognitive function in middle age as a predictor of later life health: Analyses of data from the Air Force Health Study	Data only
Boekelheide[c]	Brown University	Identifying epigenetic molecular markers of dioxin exposure in Vietnam veterans	Data and biospecimens
Chambers[c]	Mississippi State University	A longitudinal study of paraoxonase 1 (PON1) in relationship to type 2 diabetes and aging	Data and biospecimens
Haws	ToxStrategies, Inc.	Exposure–response relationship for dioxin and cancer and noncancer health outcomes in the Air Force Health Study cohort using physiologically based pharmacokinetic modeling of exposure and updated mortality	Data only
Knafl	University of North Carolina at Chapel Hill	Effects of dioxin exposure for male Air Force Vietnam veterans on reproductive outcomes	Data only
Mandel	Exponent, Inc.	The reanalysis of the Ranch Hand data	Data only
Mazur	Syracuse University	Testosterone changes	Data only
Mitchell	Emory University and Atlanta VA Medical Center	Identifying novel biomarkers of vulnerable coronary artery disease: The Air Force Health Study	Data and biospecimens
Ramos[c]	University of Louisville	Detection of L1 protein in Ranch Hand biospecimens	Data and biospecimens
Ross[c]	Pacific Health Research and Education Institute, VA	Parkinson's disease and pre-motor features of Parkinson's disease in the Air Force Health Study	Data only
Roth[c]	VA San Diego Healthcare System	Caveolin's role during healthy aging	Data and biospecimens
Seldin[c]	Boston University	Incidence of abnormal free light chains and other markers of light chain amyloidosis in veterans exposed to Agent Orange: A pilot study	Data and biospecimens
Shim[c]	Centers for Disease Control and Prevention	Monoclonal gammopathy of undetermined significance (MGUS) and microRNAs in Ranch Hand veterans	Data and biospecimens

[a] This list represents studies approved as of February 1, 2015.

[b] The date that a proposal was approved and the date the study was able to start varied depending on the type and extensiveness of the requested assets. For example, studies that requested more biospecimens (for example, more than 100) took longer to process than those requesting fewer samples.

[c] Denotes studies that were awarded pilot funding.

the AFHS assets were valuable for their research and any suggestions they could provide to make the AFHS assets research program more beneficial. Questions were related to the overall application process, quality and completeness of the data received, quality and usefulness of the biospecimens, and future plans regarding the results and additional uses of the AFHS assets. The responses were supportive of both the AFHS assets as well as the application process. Investigators who received and used data stated that it was generally complete. While at least half of the investigators' studies were still in progress when they provided feedback to the committee, they were confident that the quality of the data, both provided and generated, would allow them to draw reliable conclusions. For the studies that used the biospecimens, all of the investigators that had received biospecimens reported that the samples were in good condition, and they were able to perform the proposed assays successfully. Both DNA and microRNA were isolated successfully from the biospecimens.

CONSIDERATIONS FOR CUSTODIANS OR STEWARDS OF THE AFHS ASSETS

MFUA's experience as custodian has yielded a number of lessons learned that need to be considered and accounted for to ensure continued successful use of the AFHS data and biospecimens for new and innovative research, no matter who exercises future responsibility for the assets. These include storage (including data security), data curation, and privacy protection; laws, rules, and policies related to asset management and dissemination; and sustained funding to support these activities. Both data and biospecimens require continued management based on best practices that will also serve to make these assets more accessible and useful to the broader research community.

Data must be stored in a manner that maintains its physical and logical integrity and allows it to be easily retrievable. Data curation refers to the active management of data required to maintain it long term for preservation and reuse. Proper data documentation, quality, and accessibility are key to enhancing and enabling use internally as well as by the broader scientific community. Because of the longitudinal nature of the AFHS and the diverse manners by which data were often collected, recorded, and documented, as well as the different statistical methods used to analyze and account for missing data and loss to follow-up, the structure of the data is complex. In addition to continued storage and data curation, the biospecimens also require continued management to ensure their viability and value for use in future research.

The extensive data collected on the AFHS participants and the linkage of that data to the biospecimens provides an extremely rich research resource. However, while those data are being made available for a broad array of research, it is critical to maintain appropriate safeguards to protect participants' privacy and confidentiality. The safeguards in place encompass not only data already available but should extend to additional data that may be generated in the future from the biospecimens or from studies that seek to collect new information on cohort participants. Balancing access to the AFHS data and biospecimens while safeguarding the privacy of the AFHS participants is essential to maintaining a successful research program. The policy of the IOM advisory committee has been to limit the information released to approved researchers to only that information required for the specific research question of interest. Variables that may lead to the identification of a participant are suppressed. The committee finds that the practices used by the IOM for maintaining privacy protection of the cohort participants are appropriate and comply with legal requirements and ethical and social norms and expectations including respecting the individuals who have voluntarily provided biospecimen samples, meeting public expectations, and supporting the efficient functioning and transparent practices of the biorepository.

The current AFHS research program should be viewed as a pilot effort. To truly realize the potential of the AFHS data and biospecimens as a research asset like the Framingham Heart Study, the Atherosclerosis Risk in Communities Study, and other longitudinal epidemiologic studies will require a more sustained and possibly greater investment of money and personnel as detailed below.

FUTURE DIRECTIONS

A number of steps have been taken to prepare the data and biospecimens for use in research since the IOM began its AFHS Assets Research Program. However, the time period for evaluating the use of the data and biospecimens for other purposes has been short, and owing to the level and length of funding this pilot effort could not have the impact of a more established research program.

The longitudinal nature of the AFHS—with its extended follow-up, high rates of retention, and repeat biological samples—provides a valuable opportunity for research beyond the original aims of the study. Unused biospecimens and already-collected medical records can be used to conduct research studies in a far faster and more cost-effective manner than those that require investigators to identify and assemble such materials from scratch. Potential new areas for the AFHS assets include study of the determinants of incidence or natural history of common conditions (for example, hypertension, diabetes, or obesity) as well as the correlates of health in aging.

As the research program moves forward, short- and long-term metrics will need to be developed to evaluate its reach, value, and success. Different metrics may be appropriate for the data and biospecimens.

FINDINGS, CONCLUSIONS, AND RECOMMENDATIONS

The committee's findings fall into two overarching categories: the scientific value of the AFHS data and biospecimens and the lessons learned in managing access to the assets. These and the conclusions that flow from them are summarized in Box S-3 and explicated in Chapter 5.

When the *Disposition of the Air Force Health Study* report was published, it was presumed that a successful research program could be implemented, but there

BOX S-3
Summary of the Committee's Findings and Conclusions

Findings—Scientific Value

- A great majority of the AFHS cohort members continue to allow use of their data and biospecimens for research.
- The biochemical integrity of the AFHS biospecimens appears to be well preserved, and the biospecimens are amenable to analysis by long-established and newly developed assays.
- A broad spectrum of the scientific community has demonstrated interest in performing research on the AFHS data and biospecimens.

Findings—Assets Management

- The IOM staff have begun the process of transforming and standardizing the AFHS data into a form that is more amenable to modern data curation and research techniques.
- The IOM's data management practices have preserved the security of the AFHS data and biospecimens and the privacy of the participants' information.
- The Air Force Research Laboratory's management of the AFHS biospecimens has been successful.
- Pilot funding, provided by VA under congressional direction, catalyzed research involving biospecimens. However, it is premature to draw any conclusions about the effectiveness of the pilot funding effort.
- The IOM has created knowledge about the AFHS assets management process that will benefit future efforts.

Conclusions

- It is possible to manage the AFHS assets and perform high-quality scientific research with them.
- Sustaining access to the AFHS biospecimens and data repository benefits the veterans community and the public at large, who gain from the information derived from studies of the assets.
- The AFHS assets have been underutilized, and the custodian should continue to seek ways to improve management approaches to maximize the use of this resource in research.
- It is feasible and advisable to maintain the AFHS data and biospecimens and make them available for continued use in research.

was no way of knowing whether it would work as that committee had envisioned. That presumption has now been confirmed and the once-speculative potential of the program has been largely realized. The assets have generated continuous interest from researchers, as evidence by the more than 80 inquiries and requests between 2012 and 2014, something that is especially encouraging considering that it has not been possible to actively promote their availability until recently. Therefore, the committee concluded that it is possible to manage the AFHS assets and perform high-quality scientific research with them. Furthermore, it is feasible and advisable to maintain the AFHS data and biospecimens and make them available for continued use in research. However, while the committee is pleased that a broad spectrum of investigators is carrying out a variety of studies using the AFHS assets, it believes that the full potential of the assets has not come close to being realized.

The committee concludes that sustaining access to the AFHS biospecimens and data benefits the veterans community and the public at large, who will gain from the information derived from studies of the assets. It is too soon to know whether and which studies of the assets will yield information that will promote health and wellness in veterans or in the greater population, but the research performed to date has the potential to contribute to the current body of medical knowledge, especially the results of studies that are examining potential biomarkers of disease. This area of research has greatly advanced in recent years, allowing for studies that were not possible when the original AFHS was conducted. The combination of longitudinally collected data and biospecimens is well suited to biomarker research, which could advance the science of disease diagnosis and treatment.

Despite the fact that the AFHS is among the few long-term, prospective, epidemiologic studies that included longitudinally collected biospecimens, it has only recently been accessible to the broad scientific community for research. Therefore, there has been limited awareness and use of the assets for novel research. Publication of the results of currently ongoing research should help to address this problem, but a more concerted effort to raise the assets' visibility is also needed. Any attempts to expand the number of researchers, though, will have to be accompanied by plans for how to fulfill requests for the assets and fund that work.

The congressionally directed transfer of the AFHS assets to the custodianship of the IOM has resulted in promising research that is yielding further benefits from the government's approximately $143 million investment in the original study. Many of the data files have now been transformed and documented in a manner that facilitates research by outside investigators, and sustaining access to them along with the biospecimens will advance knowledge regarding the determinants of health and wellness.

The value of the AFHS assets lies in their combination of longitudinal epidemiologic data and associated biospecimens and would be greatly diminished if the participants' serum, whole blood, urine, semen, and adipose tissue samples are not maintained in addition to their health data and other personal information. Both resources should thus be preserved, and the funding to maintain the resources must include support for both if the assets are to remain valuable for research.

Based on these findings and conclusions,

The committee recommends that Congress continue to support the maintenance of the Air Force Health Study data and biospecimens as a resource for research and to facilitate making them available to the scientific community as broadly as possible.

This must be done in a manner that continues to preserve the privacy of study participants and the security of their data.

The committee identified two options regarding the continued maintenance and management of the AFHS assets:

1. Retain custody of the assets within the IOM's MFUA, either maintaining the current management structure—where MFUA curates the assets—or forming a partnership between the IOM and another organization to manage and distribute the data to researchers (as is presently done with the biospecimens); or
2. Transfer custody of the AFHS data and biospecimens to another organization.

The *Disposition of the Air Force Health Study* report identified several characteristics of a good custodian for the AFHS assets. Those characteristics included a demonstrated ability to properly house, secure, and manage both very large and complex data and the accompanying biospecimens (IOM, 2006). The committee finds that the IOM has met each of these criteria and has maintained and managed the AFHS assets in accordance with best practices.

MFUA has much experience in storing, administering, and disseminating epidemiologic data, and maintaining the current structure of the assets management program is a viable option. However, joining forces with an organization that administers databases or funds research as its primary mission may allow for better promotion and dissemination of the AFHS data and biospecimens for new and original research than could be accomplished by the IOM alone. This is not meant to cast doubt on the ability of the IOM to manage this program, but instead to suggest a possible alternative approach that combines the strengths of the IOM in scientific review with the institutional capabilities of a dedicated data archive. In this scenario, MFUA would retain custodianship and partner with an organization that has the infrastructure for storing, curating, and maintaining the electronic data, and experience in promoting the use of large datasets. Such a partnership would be similar to the one currently in place for the storage and maintenance of the biospecimens between the IOM and AFRL. Under such an arrangement, the IOM would retain custody of the assets and approve the type and amount of information given to researchers through an advisory committee. The partnering organization would physically hold the assets; have responsibility for cleaning and applying quality control, managing, and maintaining them; and preparing and distributing data (and biospecimen samples) to researchers as directed by the IOM. Affiliating with a clearinghouse for data and biospecimens could significantly raise the visibility of the AFHS assets and spur their greater use.

The committee does not have any recommendations concerning partnering organizations, but it can offer some examples for consideration should this option be pursued. The National Institute on Aging sponsors the Chicago Core for Biomarkers in Population-Based Aging Research, which—among other activities—maintains a directory of longitudinal studies that include biomarker and personal health data that are made available to investigators. The National Heart, Lung, and Blood Institute's Biologic Specimen and Data Repositories Information Coordinating Center serves a similar function for studies that hold assets relevant to the institute's mission. And the Inter-university Consortium for Political and Social Research "stores, curates, and provides access to scientific data so others can reuse the data and validate research findings" (ICPSR, 2014). The organization's vast collection of databases, which are primarily focused on health (including aging) and social science issues, includes some that address health questions and a small number with associated biospecimens. All are made available to interested investigators according to the rules set out by the data depositor. A thorough evaluation of partnering options would require discussions with candidate organizations to determine their willingness and capacity to take on curation of the AFHS data and a determination of the logistics and costs of preparing them for the relocation—tasks that are beyond the scope of this committee.

If Congress chooses to designate an alternate custodian to manage the AFHS data and biospecimens and to administer a research program, such a decision should be carefully considered and based on thorough comparison of the attributes and characteristics between the IOM and alternatives it evaluates. Many issues would have to be resolved before such a transfer could take place. A new custodian would need to exercise the same commitment to protecting the confidentiality of the assets as has been practiced to date; and to adhere to best practices, legal and regulatory requirements for the storage, maintenance, and dissemination of data and biospecimens. The cohort would have to consent to having responsibility for their study materials transferred to a new custodian, which based on previous experience would likely result in a further diminution of the cohort. Loss of additional cohort members will reduce the value of these assets for use in additional research because the sample size may not be adequate for many types of studies. Additionally, if the AFHS assets were transferred to a different custodian, their availability for continued use in research would likely suffer a setback because the new custodian would have to set up a new research program or integrate the AFHS into an existing one. It is unclear in what format the data would be transferred and whether the custodian would choose to create a new database using the original text files or would try to incorporate the work MFUA has done (if the systems are compatible). Finally, a new custodian would need to find an appropriate and secure location to house the freezers in which the biospecimens are stored, form a relationship with AFRL, or transfer them to a different ultralow temperature storage facility. Funding would need to be provided to support the staff needed to prepare assets for the transfer and assure that it was successfully completed.

No matter who serves as custodian of the AFHS assets, a multiyear commitment to funding this organization is highly desirable as this would permit the long-term planning and sustaining of the infrastructure needed to conduct an effective assets dissemination program.

To derive the maximum value from research conducted on the assets, steps should be taken to make sure the knowledge developed from them is widely disseminated through peer-reviewed publications and other professional venues and integrated into the database for subsequent use by other investigators. Data use agreements with investigators should continue to require that newly developed data be returned to the custodian, with a reasonable amount of time allotted for publications to first be generated. After these data are integrated, their availability should be promoted. When investigators establish that the biospecimens are amenable to analysis by particular assays, this information should be made generally available so other potential researchers can factor it into their plans.

Finally, the committee believes it important to increase the breadth of scientific disciplines that are aware of and use the AFHS assets. These data and biospecimens have been and continue to be used to make important contributions to the understanding of the toxicologic mechanisms and health impacts of exposure to herbicides and their contaminants. However, there are many additional opportunities to learn from these assets beyond the purpose of their initial collection or to use technology and science that had yet to be developed when the original study ended. Given the rapid development of new analysis techniques and instrument technologies; assay methods with improved analytical sensitivity and specificity; and an abundance of existing proteomic, metabolomic, and genetic methods that can be applied to specimens and data from the AFHS, the potential value and utility of the AFHS specimens is far greater than what was originally imagined at the beginning of the study. It will be up to the custodian to identify new areas of research and to find ways to reach the people who conduct it so they can start taking advantage of this rare and valuable resource.

REFERENCES

ICPSR (Inter-university Consortium for Political and Social Research). 2014. Elements of a public access plan for scientific data. *ICSPR data management and curation.* http://www.icpsr.umich.edu/icpsrweb/content/datamanagement/ostp.html (accessed September 22, 2014).

IOM (Institute of Medicine). 2006. *Disposition of the Air Force Health Study.* Washington, DC: The National Academies Press.

1

Introduction

Longitudinal prospective studies that follow a population for an extended period (a decade or more) are the gold standard of observational epidemiologic studies. Such studies are uncommon though, because of the great expense and time needed to conduct them. Rarer still are well-designed cohort studies that include the collection and storage of biospecimens for use in current and future analyses. The Nurses' Health Study (Colditz et al., 1997), the Framingham Heart Study (Oppenheimer, 2005, 2010), and the Atherosclerosis Risk in Communities Study (1989) are examples of such efforts that had been designed to study a particular health outcome (cancer for the former and heart disease for the two latter studies) but which have produced broader research findings as well. The design, data and biospecimen collection, and continued use of the assets assembled in these studies are considered to be exemplars of epidemiologic research.

One member of this select group—little known outside the veterans health research community—is the Air Force Health Study (AFHS): a longitudinal, prospective epidemiologic study of more than 2,700 men followed for approximately 20 years. This cohort participated in up to six intensive physical examinations with high rates of compliance. In addition to a complete record of the clinical measurements and observations at these exams, serum and other biological samples were obtained and preserved. Extensive questionnaires addressing health, lifestyle, and socioeconomic status were administered during each exam, and other information was obtained about the men's employment and potential sources of environmental exposures. While the team responsible for collecting these assets was disbanded at the end of the study, the data and the biospecimens were preserved and—at the direction of Congress—made generally available for research 3 years ago.

Because the AFHS has only recently been accessible to the scientific community (outside of the Air Force study personnel and a few limited collaborations) for research, the vast majority of its potential is yet to be realized.

In 2007, after the original study—which examined whether U.S. Air Force personnel involved in Operation Ranch Hand, the program responsible for herbicide spraying in Vietnam, experienced adverse health effects from their service—was completed and its final report published, Congress directed that the AFHS data and biospecimens be transferred to the custody of the Institute of Medicine (IOM). The IOM was directed to undertake a program of research using the assets to better understand the determinants of health and promote wellness in veterans and the general population. As a first step, the IOM contacted the participants and obtained consent for their data and biospecimens to be used in a broad range of future research, including topics not directly related to the aims of the original study. The IOM also developed and implemented procedures by which investigators could obtain access to the data and biospecimens. As part of the law that required the AFHS assets be made available for research through the IOM, Congress requested that the IOM advise them on the feasibility and advisability of maintaining the biospecimens and data and conducting additional research on them.

A number of peer-reviewed papers have been published and numerous reports and presentations made from analyses that used AFHS data or biospecimens, both during the study and following the end of new data collection. Findings from the AFHS have been used for many purposes to date—prominently, to improve the pharmacokinetic modeling of 2,3,7,8-tetrachlorodibenzo-*p*-dioxin (TCDD) and refine half-life estimates of TCDD in exposed populations (Michalek et al., 1996; Warner et al., 2005; Wolfe et al., 1994).[1] Other IOM committees have used findings from the AFHS to help inform conclusions regarding the health effects associated with exposure to herbicides and TCDD (IOM, 1996, 2000). However, the importance and applicability of the AFHS data and biospecimens extend far beyond research related to the study's original purpose (IOM, 2006).

BACKGROUND

Upon completion of the AFHS by a team of researchers under the direction of the Air Force, Congress passed three laws with sections relating to the disposition of its data and biospecimens. Appendix B contains the text of these sections, which are summarized below.

Section 602 of Public Law 108-183, the Veterans Benefits Act of 2003, directed the Secretary of Veterans Affairs to contract with the National Academy of Sciences to address several questions regarding the appropriate disposition of

[1] TCDD was an unintended byproduct of the synthesis of trichlorophenoxyacetic acid (2,3,5-T), a herbicide used for tactical defoliation during the Vietnam conflict. 2,3,5-T was also commonly used as an agricultural herbicide until the late 1960s.

the AFHS. Subsequently, in response to a request by the Secretary of Veterans Affairs, the IOM of the National Academy of Sciences established the Committee on the Disposition of the Air Force Health Study, which issued a detailed report that recommended that the AFHS data and biospecimens be maintained and used to foster new research and identified several possible appropriate custodians of such assets (IOM, 2006).

Based in part on the findings, conclusions, and recommendations in the 2006 IOM report, Congress passed Public Law 109-364, section 714 of which directed the Air Force to transfer custodianship of the AFHS's data and biospecimens to the Medical Follow-Up Agency (MFUA)—a unit within the IOM responsible for data-intensive epidemiologic studies—when the study formally concluded in September 2007. The Air Force was instructed to seek permission from the surviving members of the cohort for this transfer. This activity took place over the Spring and Summer of 2007, and more than 90% of the cohort members consented to the move.

Two years after Congress directed the IOM to become the custodian of the AFHS assets, Congress passed Public Law 110-389, of which section 803 directed the Department of Veterans Affairs (VA) to provide funding to MFUA to maintain and manage the AFHS assets and to make them "available as a resource for future research for the benefit of veterans and their families, and for other humanitarian purposes." MFUA was authorized to administer a program of research on the assets aimed at "understanding the determinants of health, and promoting wellness, in veterans." The law also directed that up to $250,000 per year of pilot funds be made available when appropriate to support investigations using the AFHS assets. At the end of the effort, Congress directed that a report be submitted to them that assessed the feasibility and advisability of conducting additional research on the AFHS data and biospecimens.

THE COMMITTEES' STATEMENTS OF TASK AND APPROACH

The IOM undertook two related efforts to carry out the tasks identified in section 803 of Public Law 110-389. First, in its role as custodian of the AFHS data and biospecimens, it established an advisory committee to facilitate new research using the AFHS assets and to review proposals from researchers who requested access to the data and biospecimens. Its statement of task is shown in Box 1-1. This activity is detailed in Chapter 3.

The advisory committee operationalized its statement of task by developing and implementing protocols for

- reviewing the content of prospective research proposals;
- assessing whether proposals merited access to the AFHS assets;
- governing the management and use of the biospecimens, including deciding whether and which biospecimens should be made available and how much material should be provided; and

BOX 1-1
Statement of Task—Advisory Committee

The U.S. Congress asked the Institute of Medicine to assume responsibility for managing a robust set of research assets that were produced in the course of the Air Force Health Study (also known as the Ranch Hand study), which operated from 1979 until 2006. Materials include extensive information on the health and socioeconomic status of the cohort, and serum, whole blood, adipose tissue, semen, and urine samples collected during six cycles of examinations. In its capacity as custodian of this material, the IOM will work with a committee of scientific advisors to foster the use of these data and specimens in new studies that will provide insight into factors affecting the health of veterans and the population at large, and to review research proposals.

- determining whether to award pilot funding to investigators who requested support.

Second, in 2014, to conform to institutional requirements, a consensus committee composed of the same membership was created to write a report regarding the feasibility and advisability of conducting additional research on the AFHS data and biospecimens as called for in the legislation. Its statement of task is shown in Box 1-2. While the consensus committee was not asked to evaluate the data and biospecimens dissemination program it does offer lessons learned from the program as part of its examination of the feasibility and advisability of maintaining the biospecimens.

The consensus committee held eight biweekly meetings via conference call from August through December 2014 and conducted a 1-day in-person meeting in Washington, DC, to examine the issues under its purview. It used information acquired by the advisory committee in its deliberations, including

- results of searches of relevant scientific literature on topics related to the AFHS data or biospecimens;
- annual updates from investigators in open sessions of committee meetings, during which the committee had the opportunity to question them; and
- input from interested parties, including representatives of the Ranch Hand Vietnam Association,[2] who presented information from the perspective

[2] The Ranch Hand Vietnam Association is an organization composed of members of the AFHS cohort.

> **BOX 1-2**
> **Statement of Task—Consensus Committee**
>
> A June 28, 2011, contract between the National Academies and the Department of Veterans Affairs entitled *Air Force Health Study (AFHS) Assets Research Program*—which operationalizes directives contained in Public Law 110-389, section 803 (Maintenance, Management, and Availability for Research of Assets of Air Force Health Study)—directs the IOM to "provide the final report to Congress at the end of the final period of performance outlining the feasibility and advisability of maintaining the biospecimens based on interest generated from the general scientific community and results of pilot projects and other research projects using the AFHS assets." An ad hoc committee will conduct a study and prepare a report in response to this charge. The final period of performance ends on June 30, 2015.

of AFHS participants; and VA, which attended the open sessions and addressed the committee.

As part of its approach to its task, the consensus committee developed and administered a survey to investigators who received data or biospecimens to solicit feedback on the process, completeness of the data, and quality of the biospecimens. The committee also used the findings and recommendations of the *Disposition of the Air Force Health Study* report as a guide for assessing what was then believed to be feasible and possible for a research program managing and using the AFHS assets and to compare with its now realized experience.

ORGANIZATION OF THE REPORT

The remainder of this report is organized into four chapters plus supporting appendices. Chapter 2 provides background information on the design and scope of the original AFHS, the types of extensive detailed data and biospecimens collected, and the usefulness of the AFHS assets, especially its biospecimens, for continued research. Chapter 3 describes the research effort under the IOM's custodianship, including examples of new research using both the electronic data and the biospecimens. The impact and value of the research to date and the AFHS assets' continued application to future research are also discussed. It also presents an overview of some of the practical challenges to managing the vast number and type of assets, such as their transfer, cleaning, processing, storage, and maintenance, and summarizes the proposal process created and used by the advisory committee to evaluate proposals for access to the AFHS assets. Chapter 4 examines the future of the research program including potential options to increase the availability and use of the AFHS assets while taking into account some of the associated legal requirements and ethical and social norms and expectations. Chapter 5 summarizes the committee's findings, conclusions, and recommendation regarding the disposition of the AFHS data and biospecimens, specifically the feasibility and advisability of their continued maintenance and use in new research. Appendix A provides short biographies of each committee member. Appendix B contains excerpts of the public laws regarding the National Academy of Sciences and the AFHS. Appendix C provides abstracts of all research approved by the advisory committee as of February 1, 2015.

REFERENCES

Atherosclerosis Risk in Communities Investigators. 1989. The Atherosclerosis Risk in Communities (ARIC) Study: Design and objectives. *American Journal of Epidemiology* 129(4):687-702.

Colditz, G. A., J. E. Manson, and S. E. Hankinson. 1997. The Nurses' Health Study: 20-year contribution to the understanding of health among women. *Journal of Women's Health* 6(1):49-62.

IOM (Institute of Medicine). 1996. *Veterans and Agent Orange: Update 1996*. Washington, DC: National Academy Press.

IOM. 2000. *Veterans and Agent Orange: Herbicide/dioxin exposure and type 2 diabetes*. Washington, DC: National Academy Press.

IOM. 2006. *Disposition of the Air Force Health Study*. Washington, DC: The National Academies Press.

Michalek, J. E., J. L. Pirkle, S. P. Caudill, R. C. Tripathi, D. G. Patterson, and L. L. Needham. 1996. Pharmacokinetics of TCDD in veterans of Operation Ranch Hand: 10-year follow-up. *Journal of Toxicology and Environmental Health* 47:209-220.

Oppenheimer, G. M. 2005. Becoming the Framingham Study 1947–1950. *American Journal of Public Health* 95(4):602-610.

Oppenheimer, G. M. 2010. Framingham Heart Study: The first 20 years. *Progress in Cardiovascular Disease* 53(1):55-61.

Warner, M., B. Eskenazi, D. G. Patterson, G. Clark, W. E. Turner, L. Bonsignore, P. Mocarelli, and P. M. Gerthoux. 2005. Dioxin-like TEQ of women from the Seveso, Italy, area by ID-HRGC/HRMS and CALUX. *Journal of Exposure Analysis and Environmental Epidemiology* 15(4):310-318.

Wolfe, W. H., J. E. Michalek, J. C. Miner, J. L. Pirkle, S. P. Caudill, D. G. Patterson, Jr., and L. L. Needham. 1994. Determinants of TCDD half-life in veterans of Operation Ranch Hand. *Journal of Toxicology and Environmental Health* 41(4):481-488.

2

The Air Force Health Study

The Air Force Health Study (AFHS) was a longitudinal, prospective epidemiologic study initiated by the U.S. Air Force in 1979. Its original purpose was to determine whether U.S. Air Force personnel (all men) who had participated in Operation Ranch Hand—the program responsible for herbicide spraying in Vietnam—had experienced adverse health outcomes as a result of their service. The study protocol consisted of three components: a retrospective mortality study, a retrospective morbidity study, and a 20-year prospective follow-up study. The most resource intensive of the components, as well as the basis for the current research initiative, was the prospective follow-up study, which consisted of six comprehensive exams. A comprehensive set of clinical measurements and observations were made and biological specimens (serum, whole blood, urine, semen, and adipose tissue) were obtained and preserved at each of the six exams, or *cycles*. Extensive health questionnaires addressing health, lifestyle, socioeconomic status, employment, and other potential sources of environmental exposures were administered during each exam cycle. Other information was obtained from personal medical records and from the participants' families. Data collection and analysis for the formal study was completed in 2006. However, due to a congressional directive, the extensive health data and accompanying biospecimens have to date remained available as a resource for additional research through the Institute of Medicine (IOM).

This chapter summarizes information on the AFHS relevant to the consideration of its further disposition. It begins with an overview of the design of the study. A detailed description of the data and biospecimens collected during the course of the AFHS that remain useful for conducting new research studies is then provided. The chapter ends with a summary of the findings, conclusions, and recommendations made in the 2006 *Disposition of the Air Force Health Study* report, the effort that laid the foundation for the current research program.

DEVELOPMENT AND DESIGN

The protocol for the AHFS was developed by a team of principal investigators and other researchers from the Air Force School of Aerospace Medicine at Brooks Air Force Base in San Antonio, Texas. It had three major components: a retrospective mortality study, a retrospective morbidity study, and a 20-year prospective morbidity follow-up study.[1] The protocol was extensively and independently reviewed, revised, and rereviewed by the Air Force and academic and scientific institutions, including the National Research Council of the National Academy of Sciences, the University of Texas School of Public Health in Houston, the Air Force Scientific Advisory Board, and the Armed Forces Epidemiological Board until the 11th version was implemented in 1982 (AFHS, 1982, 1984a). Beginning in 1980, the Advisory Committee on Special Studies Relating to the Possible Long-Term Health Effects of Phenoxy Herbicides and Contaminants—also known as the Ranch Hand Advisory Committee—was convened as the entity responsible for overseeing and monitoring the conduct of the study.

The prospective morbidity study collected data through comprehensive physical examinations (including laboratory work and biospecimen draws), participant questionnaires, and reviews of personal medical records. These in-depth information collection efforts took place at six time points in years 1, 3, 5, 10, 15, and 20 of the study. Figure 2-1 depicts the study timeline including the time of exposure to herbicides, duration of the AFHS from protocol development to termination, and IOM custodianship. In all, 2,758 individuals participated in at least one exam cycle (IOM, 2006).

Ranch Hand personnel—1,242[2]—were identified and individually matched on age, race, and military occupation (differentiated into five categories: officer/pilot, officer/navigator, officer/other, enlisted/flight engineer, and enlisted/other) to comparison participants for the first exam cycle (1982) (AFHS, 1984a). The comparison participants served in the U.S. Air Force between 1962 and 1971, were stationed in and flew cargo operations elsewhere in Southeast Asia during the Vietnam conflict, had not been exposed to tactical herbicides, and were assumed to be similar to the Ranch Hands regarding lifestyle, training profiles, and socioeconomic factors. Each Ranch Hand was matched to a pool of 8–10

[1] The IOM report *Disposition of the Air Force Health Study* provides a more detailed explanation of the design and results from the mortality and retrospective morbidity study components (IOM, 2006).

[2] The exact number of Ranch Hands varies between published reports, but 1,242 reflects the number who had served in Vietnam and were not killed in action; however, not all of them ultimately participated in the Air Force Health Study (deceased before 1982, unlocatable, or refused) (IOM, 2006).

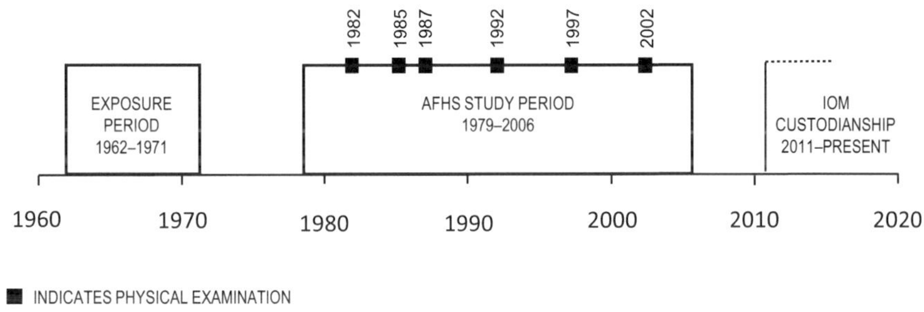

FIGURE 2-1 AFHS timeline.

comparisons,[3] who were selected based on the first living and compliant person randomly selected from the individual-level pool. Individual comparison participants remained associated with their matched Ranch Hand for the duration of the study, but those who died, dropped out, or were lost to follow-up were replaced with the next best comparable control who was living and agreed to participate (AFHS, 1982). Although not representative of the U.S. population, the sample was diverse in terms of socioeconomic status and educational background.

Questionnaires were administered and physical examinations were conducted by contract personnel who were blinded to the herbicide exposure status of participants (AFHS, 1982, 1984a, 2005). The questionnaire was developed by the National Opinion Research Center in cooperation with the principal investigators and included questions on a broad range of health effects as identified by studies of humans and animals exposed to phenoxy herbicides and dioxin, hypothetical health effects based on biochemical and biological systems, and veterans' complaints and public perception of health effects to the herbicides used (AFHS, 1984a). Portions of previously field-tested questionnaires were also incorporated in the questionnaire to maximize their validity (AFHS, 1984a). Baseline questionnaires were administered by trained professionals of Louis Harris and Associates, Inc., in participants' homes; spouses were also interviewed. In subsequent cycles, under subcontract to Science Applications International Corporation (SAIC),[4] the National Opinion Research Center administered all questionnaires, including the baseline questionnaire for new participants. In-home interviews to complete the baseline questionnaire for new participants continued through part

[3] The committee notes that the comparison cohort was selected to be comparable to the Ranch Hands, but was not truly matched in an epidemiologic sense because the follow-up time was not identical for Ranch Hands and their comparison subjects if the comparison was replaced over the duration of the study.

[4] SAIC was a contract service provider to the AFHS for all exam cycles. In cycles 2–6 SAIC administered the questionnaires and conducted the physical exams at the Scripps Clinic and Research Foundation. SAIC created and documented the protocols for collection, shipment, assay, and storage for the biological specimens for cycles 2–6 and created data dictionaries for later cycles.

of cycle 3 (1987), after which all questionnaires were administered at the physical exam site (AFHS, 2005). New participants—primarily replacement comparison participants—received the same questionnaire that was administered at the cycle 1 evaluation as well as the current cycle questionnaire (AFHS, 2005). An interval questionnaire was developed and updated for each cycle to capture new information and update existing information for all returning participants. Questions in the follow-up cycles were formulated to cover the time interval since the last questionnaire (Robinson, 2007). When a question was added to the questionnaire for a particular follow-up cycle, it was included in subsequent questionnaires to establish a longitudinal record for the item (AFHS, 2005).

All study participants who had completed the questionnaire were invited to complete a 2.5-day physical examination. Persons who refused to complete a questionnaire were excluded from the physical exam. If the person who refused were in the comparison group, a compliant replacement participant was selected from among the other comparisons in the matched set. Examiners were required to strictly adhere to the handbook included in the protocol to ensure high-quality data collection and to minimize variability. All laboratory tests were also subject to strict protocols and quality control (AFHS, 1984a). Participants were asked to bring a copy of their medical and dental records to the physical exam. These records were reviewed and information from them extracted and added to the AFHS repository. As with the questionnaire, the physical examination included health outcome measures and endpoints that were known or suspected to be affected by phenoxy herbicides and dioxin exposure based on results from scientific literature reviews (AFHS, 1984a). The final version included those outcome measures and endpoints that were feasible, practical, and of limited invasiveness (IOM, 2006). The baseline physical examinations were conducted by Kelsey-Seybold Clinic, P.A., in Houston, Texas. All physical examinations for cycles 2–6 were administered by Scripps Clinic and Research Foundation in La Jolla, California. It took approximately 10 months to conduct physical exams for all participants (Cecil, 1986).

ASSETS COLLECTED

The AFHS assets consist of electronic data from questionnaires, physical exams, and other sources as well as frozen laboratory biospecimens. The following section details the voluminous amounts of health and other types of data collected.

Electronic Data

Information was gathered through comprehensive physical examinations, questionnaires, and reviews of medical records. Box 2-1 summarizes the major types of information collected. At the end of the AFHS in 2006, more than 123,000

BOX 2-1
Summary of Clinical, Laboratory, and Questionnaire Information Collected by the Air Force Health Study

Biospecimens: serum, whole blood, adipose tissue, semen, urine

Clinical measures: spirometry, chest radiographs, electrocardiograms, psychological testing, dermatology exams, peripheral vascular exams, neurological assessments

Demographics: education, employment, income, military experience, age, race

Endpoints by organ system: neurologic, endocrine, pulmonary, immunologic, renal, gastrointestinal, hepatic, hematologic, dermatologic, psychiatric, neoplasia, cardiovascular, reproductive

Familial factors: marital and fertility history, child and family health

General health: health habits, physical activities, leisure activities, toxic substances exposures, vital status, cause of death

electronic files were stored on AFHS servers and later transferred to the IOM. These files were stored as text files, images of X-rays and electrocardiograms, analysis files (Statistical Analysis System databases and programs, Fortran programs, Excel spreadsheets), PDFs of questionnaires and physical exam forms, and codes, including translation values to those used in data files. Content of the files includes questionnaires and physical exams from all six cycles, some of the analysis files used to create the AFHS reports for each cycle, diagnostic codes, dioxin analyses for cycle 3, dioxin data for cycles 3–6, last known addresses of all cohort members updated in 2007, journal articles and AFHS technical reports, master files to link personally identifiable information to case numbers, medication codes, the questionnaire on exposure to herbicides administered to Ranch Hand ground crew, mortality information updated in 2007, tour of duty histories, and freezer location of specimens (Robinson, 2007). About 25,000 of transferred files were considered high priority by the advisory committee and the IOM data staff and required analysis and processing for the creation of the master data files. These master data files were needed so data and samples could be provided to researchers. Each exam cycle includes its own master data files, which are used as the basis for all analyses, and within a cycle, there are separate files for different components of the cycle's data gathering effort. Data in these files are stored by the participant's unique identification number. Preparation of data files continued throughout the entire duration of the research program, with more than 50,000 files processed as of November 1, 2014.

Questionnaire data included information relating to demographics; employment; child and family health; health habits; recreation, leisure, and physical activities; toxic exposures; military experience; and wartime herbicide exposure. Data collected during physical examinations comprise indices of health status that encompassed general health and endpoints by major organ system. Appendix B of the *Disposition of the Air Force Health Study* report provides an extensive summary of the health outcomes and endpoints collected in the course of the AFHS by cycle and organ system, including the more than 200 laboratory tests performed on the biospecimens. Table 2-1 provides an overview of the number of consented participants that completed the questionnaire and physical exam in each examination cycle; those data are available for use in research.

Additional sources of data collected in the course of the AFHS included medical records from the participants' physicians, dentists, and other health providers; vital status records, such as birth and death certificates; and military administrative records that contained duty station orders, flight records, performance reports, awards and decorations, and discharge documents (AFHS, 1984a). These data and other exam materials such as electrocardiogram strips, chest radiographs, and high-resolution dental videos were scanned and stored as image files in irregular formats. All of these files were transferred to the IOM when it became the custodian of the AFHS assets and have the potential to be used in research; however, the IOM has been unable to extract and incorporate those data into a usable form (see Chapter 4).

Another component of the AFHS focused on reproductive outcomes because the possibility of increased risk of birth defects in children of Vietnam veterans was of concern. Reproductive outcome information was collected in all six cycles. In addition to the participant questionnaires that included questions on marital and reproductive history, data were gathered and compiled from interviews with current and former spouses and partners conducted in cycles 1–4. Permission

TABLE 2-1 Number of Consented Participants Who Completed Questionnaires and Physical Exams by Year Collected

Component	1982 Cycle 1	1985 Cycle 2	1987 Cycle 3	1992 Cycle 4	1997 Cycle 5	2002 Cycle 6	Completed all six cycles
Questionnaire	1,983	1,884	1,869	1,816	1,711	1,569	1,269
Ranch Hand	908	851	839	800	726	645	577
Comparison	1,075	1,033	1,030	1,016	985	924	692
Physical Exam	1,846	1,884	1,869	1,817	1,711	1,571	1,214
Ranch Hand	866	851	839	800	726	645	561
Comparison	980	1,033	1,030	1,017	985	926	653

for the release of medical data and records related to the participant, spouse, and children was also obtained. In total, data were collected on 9,921 conceptions and 8,100 live births. Other information collected included fertility, miscarriages, stillbirths, induced abortions, intrauterine growth retardation, birth defects, and general health and well-being of children through age 18 (IOM, 2006; Michalek et al., 1998). These analyses were limited to those participants for whom serum dioxin was measured in the 1987 or 1992 exam cycles (IOM, 2006).

Although some participants (233 Ranch Hands[5]) were considered exposed to high levels of military herbicides, most were not. Some of the men were exposed to elevated levels of dioxin, but the comparison participants and even most of the group that was potentially exposed to herbicides had serum dioxin levels well within population background levels.[6] Herbicide exposure was approximated through a series of serum dioxin assays performed in cycles 3–6, but AFHS reports warned that this qualitative exposure index was "not a good measure of actual dioxin exposure" (AFHS, 1991; IOM, 2006, p. 42). Comparison participants were considered to have background levels of dioxin (defined as ≤ 10 ppt), and Ranch Hands with levels unknown or ≤ 10 ppt were excluded from analyses. Pavuk and colleagues (2014) measured serum concentrations of 2,3,7,8-tetrachlorodibenzo-*p*-dioxin (TCDD) and other dioxin-like compounds in samples from 777 Ranch Hands and 1,173 comparisons who participated in the 2002 exam cycle. The total dioxin equivalents in Ranch Hands (with the exception of dibenzo-*p*-dioxin congeners) were not statistically different from those of the comparisons or those reported for similarly aged males in the 2001–2002 National Health and Nutrition Examination Survey results. These findings provide empirical evidence that apart from TCDD, Ranch Hands were not more highly exposed to dioxin-like compounds than were the comparisons or the general U.S. population. Should potential exposure to dioxin be considered a confounder for certain research questions, the cohort could be restricted to comparisons participants only or all participants with dioxin levels equal to or below background levels.

The AFHS data represents an older cohort of participants, significant numbers of whom have been diagnosed with several types of conditions or diseases, and some who have died of various causes. Table 2-2 shows the numbers of unique Ranch Hand and comparison participants that have *International Classification of Diseases, Ninth Revision* (ICD-9) diagnoses by system category. For example, a person with ischemic heart disease and diabetes will be counted in both the

[5] Determined by a compilation of different dioxin measures used by the original AFHS custodians where dioxin level was > 90 ppt.

[6] Dioxins and dioxin-like compounds are a broad class of chemical compounds that persist in the environment and are found to be present in human tissues at low concentrations primarily through ingestion of meat and other animal products (CDC, 2009). Other compounds enter the body through inhalation exposures, some of which are toxic, such as TCDD, a contaminant of the Agent Orange herbicide used in Vietnam. Toxic equivalency has been shown to be affected by age, gender, race, smoking status, and percent body fat (Ferriby et al., 2007).

TABLE 2-2 Diagnoses by System for Ranch Hand and Comparison Participants Who Consented to Having Their Data and Biospecimens Used in New Research Studies

Diagnosis Category (ICD-9 code)	Ranch Hands N = 922	Comparisons N = 1,288	Total N = 2,210
Circulatory disease (390–459)	837	1,155	1,992
Respiratory disease (460–519)	686	935	1,621
Digestive disease (520–579)	455	582	1,037
Cancer (140–239)	591	814	1,405
Endocrine disease (240–279)	725	1,012	1,737
Nervous system disease (320–359)	432	565	997
Genitourinary disease (580–629)	677	910	1,587
Infectious or parasitic disease (001–139)	661	879	1,540
Mental disorders (290–319)	692	915	1,607

circulatory disease and the endocrine disease categories. However, if a person has been diagnosed with more than one condition in a category, they are only counted once in that category. The counts include both deceased and living individuals.

Cause-specific and all-cause mortality comparison studies of Ranch Hands and their comparisons have been made throughout the original AFHS and updated at a few points since its termination (AFHS, 1984b; Pavuk et al., 2006). Mortality data was provided to the IOM for 460 consented Ranch Hand and comparison participants. Since the previous update in 2003, an additional 155 Ranch Hands and 188 comparisons have died. In all, information and biospecimens are available on 2,210 Ranch Hand and comparison participants, including 757 deceased participants, that can be used in new research studies. Updated mortality information was requested through 2012 (the last complete year available).

Using mortality information updated through December 31, 2003, Pavuk and colleagues reported on a cohort of 20,343 veterans (1,263 Ranch Hands and the pool of 19,080 comparable participants) who were followed after their service in Vietnam. After adjusting for year of birth and military occupation, the authors found the all-cause relative risk (RR) of death was significantly increased (RR = 1.25, 95% CI: 1.1, 1.4) in Ranch Hands relative to comparison participants (Pavuk et al., 2006). The increased death rate was mostly attributable to increases from diseases of the circulatory system (RR = 1.4, 95% CI: 1.1, 1.8). The RR of death from cancer was not significantly increased. The number of deaths from other categories of diseases, including respiratory, endocrine, and digestive diseases, was too small for relative risks to reach statistical significance but risk appeared to be somewhat elevated among AFHS participants. Stratifying participants by dioxin exposure level did not result in any significant increases in the risk of death overall for all causes or by cancer in the cohort. However, deaths from circulatory diseases showed a significant increasing trend for level of dioxin exposure and risk of death caused by circulatory diseases.

Biospecimens

The AFHS repository contains more than 91,000 unaliquoted biospecimens.[7] The number of samples varies by type of specimen, participant, and cycle. These were collected from study participants at each exam cycle and preserved to be used for future analyses. In the course of the AFHS, more than 200 clinical laboratory tests and measures were conducted and evaluated, and more than 60 of these were measured at all six cycles, such as red and white blood cell counts, hemoglobin, platelet count, total testosterone, bilirubin, cholesterol, and glucose. The number and type of laboratory tests performed at each physical examination changed over time reflecting changes in science and technology. Tests included blood draws, urine and semen collections, skin and fat biopsies, and stool smears; however, skin biopsies and stool smears were not retained for future research. The AFHS biospecimens are unique in that some types—such as serum and urine—were collected longitudinally across multiple cycles, while semen and whole blood were collected at a single exam cycle and adipose tissue was collected only from a subset of individuals in one exam cycle. A separate blood draw was performed to obtain samples for a TCDD (dioxin) assay carried out by the Centers for Disease Control and Prevention that was performed on 777 Ranch Hand and 1,174 matched control participants during cycles 3–6. Multiple blood samples were drawn for a subset of the population for use in dioxin biological half-life studies.

Biospecimen collection activities were conducted by contractors for all cycles. A detailed protocol is not available for cycle 1, but the protocols for collection, shipment, and storage of various biospecimens for cycles 2–6 are documented. Serum was isolated from blood samples within 2 hours of blood collection (Pavuk et al., 2007). Biospecimen aliquots were packaged on dry ice and shipped overnight to the AFHS facilities in San Antonio, Texas, where they were maintained in –70°C freezers (IOM, 2006). No samples were lost for any of the six cycles during shipment from the exam site to Brooks Air Force Base in San Antonio, Texas, or when the freezers were transferred from Brooks Air Force Base to Wright-Patterson Air Force Base in Dayton, Ohio. Because the IOM does not have an in-house capability to maintain biospecimens, the AFHS biospecimens are currently securely stored and maintained at the Air Force Research Laboratory of Wright-Patterson Air Force Base under the aegis of the IOM.

USEFULNESS OF THE BIOSPECIMEN REPOSITORY

The stored biospecimens can be linked to a wealth of information collected from the physical exams, clinical measurements, and questionnaires over time. The AFHS assets are exceptional in the sheer multitude and range of types of

---

[7] Samples collected and stored in large volumes for laboratory or chemical analysis that have not been divided into smaller volumes.

information available for each participant. In addition to the large amount of clinical information associated with this collection, there are two additional valuable features of the biospecimen repository, as identified in the 2006 report *Disposition of the Air Force Health Study*. First, the repository contains matched sample types—serum, whole blood, adipose tissue, semen, and urine—for certain cycles. Specimens collected at each examination were collected on the same day. Matched samples are valuable for research because they can be used to investigate the correlation of findings in one sample type, such as adipose tissue, with another sample type that is less invasive or more routinely obtained, such as serum.[8] This type of biological sample collection is seldom conducted despite its recognized value because of the difficulty and expense. Moreover, it is rare to find adipose and semen samples in biological sample repositories, and even rarer to find them matched with serum as in the AFHS.

Second, the repository contains serum and urine specimens collected longitudinally (in other words, repeat samples taken from participants at several different cycles) between 1982 and 2002 from a relatively large population of men. The repository contains information from 2,210 participants who participated in at least one cycle and gave consent for their materials to be used in research. Table 2-1 provides the number of individuals that completed the questionnaire and physical exam by cycle. As noted earlier, many of the same questions were asked and information on health endpoints collected at each cycle to establish a longitudinal record. Table 2-3 provides the number of participants with biospecimens available for each of the six collection time points.

Blood serum (multiple aliquots ranging between 2–10 mL per aliquot) was collected at each exam, and urine was collected at the first three exams. There are 1,204 participants (556 Ranch Hands and 648 comparisons) with blood serum samples available for all six cycles. These matched and longitudinally collected samples have been stored and maintained under conditions that permit their continued viability and use in research. Semen and serum samples collected in 1982 are 30 years old, but Pavuk (2006) and current investigators have shown that time of storage does not appear to affect their viability or properties. The specimens have been found to be suitable for many current assays and potentially for assays not yet identified. Chapter 3 has further discussion on time of storage and effect on viability.

The importance of collecting, documenting, and storing biological samples for research was not as appreciated at the time the AFHS was conducted as it is now. As noted by the National Cancer Institute,

> The lack of standardized, high-quality biospecimens is widely recognized as a significant roadblock to cancer research. . . . One of the most difficult problems that will drive 21st century cancer research [is] . . . the limited availability of carefully collected and controlled,

[8] For example, one could examine levels of various lipids in serum and markers of lipid breakdown products in matched adipose tissue collected on the same day.

> high-quality human biospecimens annotated with essential clinical data and properly consented for broad investigational use. (NCI, 2014)

In a survey of 456 U.S. biological sample repositories, only 7% were established in 1980 or earlier while 59% were established after 2001 (Henderson et al., 2013). It would be extremely difficult, if not impossible, to independently replicate the samples (let alone the voluminous amounts of accompanying detailed data) collected from a similar population over a similar time period from biological repositories in operation at the time. Given the rapid development of new analysis techniques and instrument technologies; assay methods with improved analytical sensitivity and specificity; and an abundance of existing proteomic, metabolomic, and genetic methods that can be applied to specimens and data from the AFHS,

TABLE 2-3 Number of Consented Participants with Biospecimens Available by Type and Year Collected*

Specimen Type	1982 Cycle 1	1985 Cycle 2	1987 Cycle 3	1992 Cycle 4	1997 Cycle 5	2002 Cycle 6
Blood Serum						
Ranch Hand	858	849	831	799	723	644
Comparison	975	1,027	1,023	1,011	977	924
Total	1,833	1,876	1,854	1,810	1,700	1,568
Whole Blood						
Ranch Hand						644
Comparison						923
Total						1,567
Urine						
Ranch Hand	666	840	827			
Comparison	580	1,020	1,018			
Total	1,246	1,860	1,845			
Semen						
Ranch Hand	668					
Comparison	780					
Total	1,448					
Adipose Tissue						
Ranch Hand					98	
Comparison					144	
Total					242	

* Available amounts will change as investigators use the specimens. Data are current as of November 1, 2014.

the potential value and utility of the AFHS specimens is far greater than what was originally imagined at the beginning of the study.

Not only does the AFHS repository contain matched and longitudinally collected samples, but the biospecimens are linked to data collected at each exam cycle including routine information (e.g., height, weight, blood pressure) as well as medical histories and risk factors (such as history of smoking and alcohol use or other avenues of exposure to specific chemicals), which is extremely valuable for epidemiologic research. The linked information allows researchers to stratify participants on the basis of important factors related to an outcome or disease that might not otherwise be available in other cohorts that do not have these large amounts of clinical information linked to the biospecimens. For example, researchers could investigate levels of blood glucose or other biomarkers of diabetes and risk of stroke, myocardial infarction, or other diseases in the AFHS cohort over the 20-year duration of data and specimen collection. Testing of longitudinally collected samples for the same biomarker offers an exceptional opportunity to determine whether changes in a particular biomarker predispose an individual to develop a specific disease or outcome.

Types of Biospecimens

The AFHS repository contains several types of biospecimens—some of which are held by relatively few biological sample repositories—that are valuable because they can be linked to detailed information about the individuals who participated in the study and because they were collected on the same date of the examination cycle, making it possible to determine associations between various biomarkers of exposure and outcomes of interest. The sections below address the various sample types and their research value.

Adipose Tissue

Adipose tissue is seldom found in biorepositories. Indeed, in a 2014 survey of 261 U.S. biological sample repositories, none reported adipose tissue (Edwards et al., 2014). Adipose tissue is useful for many types of epidemiologic investigations. For example, it is useful in nutritional studies because there is evidence that measurement of adipose tissue biomarkers is a more accurate measure of long-term dietary intake than dietary questionnaires because they rely on objective and measurable outcomes rather than memory, self-reported information, or interviewer-collected information that might be biased (Baylin et al., 2002). Adipose tissue has been used to quantify various biomarkers including fatty acid intakes (Baylin et al., 2002), long-term intake of carotenoids (El-Sohemy et al., 2002), and other dietary components.

Measurement of various biomarkers in adipose tissue has also been used as an indicator of cachexia (a wasting of lean and adipose mass that is often associated with cancer) (Batista et al., 2013). Adipose tissue is used to quantify exposure to lipophilic pesticides and chemicals including dioxin, and it has been proposed as a useful tissue for measuring biomarkers of cancer (particularly colorectal and breast cancer) (Campbell et al., 2009) as well as adipose tissue dysfunction that may lead to obesity (Kruijsdijk et al., 2009). Adipose tissue can be used to confirm the presence of disease, such as light chain amyloidosis (Lavatelli et al., 2008) and to examine DNA methylation marks to make comparisons of chemical exposure-related effects within the tissue.

Semen

Like adipose tissue, semen is seldom available in biological repositories; in a survey of 261 U.S. facilities, only one held semen (Edwards et al., 2014). Because it contains citric acid, amino acids, fructose, various enzymes, potassium, zinc, and various other components, it is useful for many types of epidemiological studies. The semen from a single ejaculation varies from 2–5 mL and may contain up to 600 million sperm. The viscous fluid in semen provides a nutrient-rich medium through which sperm can move and can also act as the carrier for sexually transmitted diseases.

Semen is most often used in studies of fertility, and there are many biomarkers to quantify in vivo fertility. The classic biomarker tests of fertility in semen are measures of follicle stimulating hormone, luteinizing hormone, and testosterone levels; however, new fertility biomarkers are being developed continuously and may provide equivalent results (Sabetian et al., 2010). Some seminal plasma proteins are negatively related to fertility (seminal plasma proteins 1 and 2) while others are positively related to fertility (cysteine-rich secretory protein 3) (Novak et al., 2010).

Semen biomarkers may be used in studies of prostate cancer. For example, Selth and colleagues (2014) showed that levels of microRNAs in seminal fluid were increased in prostate tumors and may help to detect prostate cancer at early stages. Semen has also been used to identify prostatitis (an inflammation of the prostate gland that affects about 10% of U.S. men). For example, one study identified 59 putative biomarkers in seminal fluid that may represent early markers of prostatitis (Kagedan et al., 2012).

Semen has been used to quantify exposure to pesticides and other chemicals with some studies suggesting that semen quality may be reduced in normally fertile men who have been exposed to pesticides and other agricultural chemical including the herbicides alachlor, atrazine, and diazinon (Swan et al., 2003). In addition, semen has been used to examine occupational exposure to polycyclic aromatic hydrocarbons. Workers highly exposed to these chemicals exhibit higher levels of sperm DNA damage and higher levels of polycyclic aromatic hydrocarbons biomarkers than nonexposed workers (Hsu et al., 2006). In addition, various biomarkers, including those for nutritional factors, oxidative damage, and antioxidants, may be measured in seminal fluid to determine nutritional status and levels of intensive exercise (Tartibian and Maleki, 2012).

Serum

Serum samples were collected at every exam cycle, and are used in a variety of standard clinical tests of physiologic conditions such as blood sugar levels and liver function, owing to the minimally invasive nature of the sample collection and the stability of many analytes during transport and processing. Serum can also be used to identify biomarkers that serve as surrogate markers of disease where direct sampling of the diseased organ is not possible. For example, serum samples can be used to test for circulatory protein biomarkers of cardiovascular disease (Battistonia et al., 2012), and a panel of multiple serum protein biomarkers was found to enable assessment of disease activity in rheumatoid arthritis (Centola et al., 2013). Other serum protein biomarkers have been characterized for early detection of colorectal cancer (Chen et al., 2014).

Serum samples can be used to analyze nonprotein biomarkers, such as metabolites (Psychogios et al., 2011), microRNAs (Condorelli et al., 2014), and circulating tumor DNA (Nguyen and Sim, 2014). MicroRNAs are known to be relatively stable in frozen serum, and microRNA profiles have been associated with different tumor types as well as a range of diseases such as cardiovascular disease, stroke, and multiple sclerosis (Reid et al., 2010). Circulating tumor DNA—which is fragmented tumor DNA that can be extracted from either serum or whole blood—can be analyzed for a variety of research purposes, including early detection and assessment of molecular heterogeneity of overall disease (Diaz and Bardelli, 2014) and tumor-specific genetic and epigenetic markers in gliomas of various grades (Lavon et al., 2010).

Urine

Unlike adipose tissue and semen, urine is a relatively common type of sample available in biological repositories because it is easily collected and useful for many types of epidemiologic studies. In their review of U.S. biological sample repositories, Edwards and colleagues (2014) found about one-third of them held urine samples. Urine is principally water and contains various organic compounds, nitrogen, salts, proteins, hormones, and other products. Typical basic urine tests include color, clarity, odor, specific gravity, pH, protein, glucose, nitrites, leukocyte esterases, ketones, blood cells, crystals, and bacterial or yeast cells and/or parasites. The type of urine test (one time or 24-hour) may also influence the tests that can be performed. Many studies collect a convenience sample of urine (taken at one time point); however, the protocol of the AFHS specified a 24-hour urine collection, of which 100 cc were aliquoted and frozen for future research (AFHS, 1984a). The 24-hour urine collection is more valuable than a convenience sample because it permits a more comprehensive examination of urine components over the course of 1 day and also permits a one-time estimation of normal liquid intake for the individual. Outcomes measured by the AFHS included volume, delta-aminolevulinic acid, coproporphyrins, uroporphyrins, porphobilinogen, and creatinine.

Urinary biomarkers have been developed to measure an array of factors related to various diseases, nutritional status, early- and late-stage cancer, and exposure to pesticides and chemicals. Such biomarkers include those that can assist in detecting early stages of diabetic kidney disease (Wang et al., 2013) and detection of multiple sclerosis (Dobson, 2012). Although urine has long been used to quantify drug use, more recent applications allow quantification of alcohol use. Ethyl glucuronide is a highly sensitive marker of alcohol consumption—it is so sensitive that individuals who recently used household products containing alcohol, but did not consume alcohol, can test positive (Zolas and Sher, 2013). In addition to drugs and alcohol, urine can often be used to quantify exposure to various pesticides and other chemicals, but measured exposure may yield different results depending whether adjustments are made by volume or by creatinine (Fortin et al., 2008). Urine is also commonly used to measure recent dietary intakes of various foods. For example, the presence of urinary 1- and 3-methylhistidine is an indicator of recent meat consumption (Cross et al., 2011). Research using urine to attempt to discover biomarkers for early detection of disease, especially cancers, continues to be an active area of study. Discovery of such noninvasive methods and measures that could accurately be used for early detection of these diseases, especially in clinical settings, would be an important advancement.

Whole Blood

Whole blood samples, which are stable during prolonged storage, were collected only in the final AFHS exam cycle (2002), and are common in biological sample repositories. Whole blood samples can be used to extract DNA that can then be analyzed to provide information on genetic markers of disease predisposition or prognosis. For example, DNA analysis of single nucleotide polymorphisms (single nucleotide polymorphisms genotyping) has been used to identify genetic markers conferring susceptibility to Crohn's disease (Yamazaki et al., 2005), age-related macular degeneration (Klein et al., 2005), and associations with environmental contaminants, such as dioxin-like compounds (Urban et al., 2011).

Continuing advances in technologies for sequencing of DNA now enables data from an entire genome to be obtained from a single sample aliquot in a short period of time. In the past 10 years, new types of sequencing instruments have been introduced that have accelerated data collection rates for DNA sequencing from years to weeks. For example, the data to decipher an entire human genome can be available within 2 weeks. Advances in technology are likely to further reduce the time needed for generating data from whole-genome sequencing, and further transform how such research is conducted (Mardis, 2013).

Extracted DNA can be used to analyze epigenetic modifications of DNA such as methylation. Epigenetic processes alter the accessibility and expression of genes without changing the DNA sequence. Methylation has important implications for normal development and disease, and thus profiling DNA methylation across the genome is important for understanding the influence of epigenetics. Technology for analyzing DNA methylation has progressed to the point where "analyses that previously were restricted to specific loci can now be performed on a genome-scale and entire methylomes can be characterized at single-base-pair resolution" (Laird, 2010, p. 191). Altered DNA methylation marks can be linked with environmental exposures and to various cell types (Hou et al., 2012; Perera and Herbstman, 2011).

Ideally, whole genome data generated by genetic and epigenetic analyses by a single investigator will also be made available to other investigators (see Chapter 4), so the samples do not need to be used to generate the whole genome data more than once. The high-throughput nature of next-generation sequencing methods will allow investigators to use the subset of the data that are relevant to their study.

FINDINGS, CONCLUSIONS, AND RECOMMENDATIONS FROM *DISPOSITION OF THE AIR FORCE HEALTH STUDY*

Before the AFHS was scheduled to end in 2006, Congress asked the IOM to convene an expert committee and conduct an assessment to address several questions regarding the disposition of the AFHS, and specifically whether there was scientific merit in retaining and maintaining the data and biospecimens. The IOM appointed a committee to address the questions, and the key findings, conclusions, and recommendations from *Disposition of the Air Force Health Study* are summarized in Box 2-2. That committee concluded that there was scientific merit in retaining and maintaining the assets after the study's scheduled termination date and recommended that further study using the AFHS data and biospecimens was advisable (IOM, 2006). Both the IOM committee and the Ranch Hand Advisory Committee, the interagency workgroup established in 1979 to oversee the AFHS, concluded that the AFHS assets have value beyond studying the effects of dioxin and herbicide exposure on health (RHAC, 2000).

The IOM committee found that the medical records, other study data, and biospecimens collected over the study's duration had been properly maintained; however, it noted that the assets were not organized and documented in a manner that allowed them to be easily understood, evaluated, managed, or analyzed by persons outside of the AFHS researchers, program staff, and personnel. Proper organization and documentation of both the data and biospecimens is necessary to maximize their effective use, especially future use by persons outside of the

BOX 2-2
Summary of Key Findings, Conclusions, and Recommendations from the *Disposition of the Air Force Health Study* Report

- There is scientific merit in retaining and maintaining the medical records, other study data, and laboratory specimens collected in the course of the Air Force Health Study (AFHS) after the study's currently scheduled termination date: no other epidemiologic dataset on Vietnam veterans contains as detailed information over as long a time period, the data appear to be of high quality and the specimens well preserved, and analysis of the assets has contributed to the literature addressing the health of Vietnam veterans.
- Obstacles exist to retaining and maintaining the AFHS data assets. These relate to factors intrinsic to the study's design, resulting from implementation decisions made by the investigators, relating to documentation and organization of the data assets, and addressing the preservation of the privacy of the study subjects and the confidentiality of their personal information. The committee believes that the identified obstacles are surmountable.
- Further study of the AFHS medical records, other study data, and laboratory specimens is advisable. This should be accomplished by making these materials available for research via a custodian that takes an active role in fostering use of the assets. Five years after the chosen custodian assumes responsibility, a committee should be convened to evaluate whether any further support should be extended to the maintenance of access to the data or the biospecimens.
- The potential value and relevance of further study of the AFHS data assets rest in the application of the results of future health research on the data assets. This research could encompass using novel analysis approaches, employing new technology and techniques, and examining data and outcomes not evaluated to date. The cost of such work will vary greatly, depending on the research question that is addressed.
- The committee cannot offer a specific recommendation on the federal or nonfederal entity best suited to continue the study of the AFHS data assets but has identified a number of options that could be pursued successfully.
- Independent oversight of future research using the AFHS data assets is advisable, and should be provided through the review of proposals for scientific merit and adherence to ethical, legal, and related considerations by an Institutional Review Board and, separately, an advisory and oversight board. Additionally, research should be carried out in a manner transparent to study subjects, through systematic communication of research plans and results.

SOURCE: IOM, 2006.

AFHS. At the time of writing that report, much of the knowledge and nuances of the data files was stored "in the collective minds and personal notes of the AFHS staff charged with maintaining it" (IOM, 2006, p. 85). As part of its recommendation to organize and document the assets prior to the scheduled termination date, the committee specified such actions as creating detailed and comprehensive data dictionaries for master data files for each examination cycle; a comprehensive dictionary of the variables contained in the master data files, organized by examination cycle and by questionnaire, physical examination report, or other data intake instrument; a master codebook that would constitute a comprehensive distillation of database contents, such as from which examination cycles a particular piece of information was gathered and the variable name(s) associated with that information over the course of the study; and a document or database describing the contents, format, and location of the AFHS collection of assets that have been scanned into PDF image files. Other items that the committee believed should be included in the comprehensive data dictionary were a synopsis of the variable names and their descriptions, the summary variables created and codes used, the number of study participants examined for each test or outcome, changes in the database structure from the previous examination, and relevant notes on data comparability between cycles. The committee recommended that these documents be in a form and format that facilitates easy access to their contents. For those variables specific to the laboratory results, it suggested that the dictionary contain descriptions of assays, units of measurement and normal ranges, and data codes. Notation should include whether any attributes of a variable have changed over the course of the study; for example, noting increased sensitivity of a biomarker where in earlier exams a participant might have tested negative and with increased sensitivity may have then tested positive in a later exam, although the value may be the same.

As is the case with the electronic data, the *Disposition of the Air Force Health Study* committee noted that obstacles, including privacy concerns, exist in maintaining the laboratory biospecimens. Privacy concerns relate to safeguarding the personally identifiable information linked to the biospecimens and any information, such as a particular diagnosis, that can be determined from the biospecimens, rather than the fact that DNA can be extracted from some of the biospecimens. (DNA, although unique for each individual, is not itself identifiable unless matched to another identified sample.)

The committee was told of a planned reassay on a selected set of biospecimens using endpoints with recorded historical values to evaluate the stability and condition of the biospecimens, but results were not available at the time the report was published. The committee recommended that the Air Force fully document and inventory the biospecimens prior to the study's scheduled termination date because such an activity would be vital to facilitating any possible future use. Recommended actions that should be part of completing an inventory of the biospecimens included verifying each biospecimen's location and ascertaining the number and volume of aliquots and type of samples; visually inspecting each specimen during the verification process to identify any problematic samples; creating a single biospecimen database that includes case number, exam cycle, specimen type, and freezer location and compiling all information regarding biospecimen history (receipt, realiquoting, freeze–thaw cycles, dispersal, etc.) into a single reference database; and compiling all protocols regarding receipt, maintenance, dispersal, and return of biospecimens for all cycles into a single reference document.

The IOM committee further concluded that concerns regarding privacy and other "ethical, legal, and social issues are not an intrinsic obstacle to retaining and maintaining the AFHS medical records, other study data, and laboratory specimens" (IOM, 2006, p. 116) and are, in fact, surmountable. Examples of these ethical, legal, and social issues include maintaining confidentiality and security of participants' identifiable health information and obtaining informed consent from the cohort participants to use the assets in broad research. However, the committee also noted that attention to such concerns and issues must be a major consideration in decisions regarding the future disposition and use of these assets. Given this, and that the committee concluded that there is scientific merit in retaining and maintaining the AFHS data and biospecimens, the committee further concluded, "Further study of the AFHS medical records, other study data, and laboratory specimens is advisable. The potential value and relevance of extending the study of the AFHS data assets rests in the application of the results of future research on the assets" (IOM, 2006, p. 116). The committee then identified several possible applications that are further explored in Chapter 3 along with the many nuances and steps between the recommendations made by the 2006 IOM committee and implementing them into a successful management program.

REFERENCES

AFHS (Air Force Health Study). 1982. *An epidemiologic investigation of health effects in Air Force personnel following exposure to herbicides: Study protocol, initial report.* SAM-TR-82-44. Brooks AFB, TX: USAF School of Aerospace Medicine.

AFHS. 1984a. *An epidemiologic investigation of health effects in Air Force personnel following exposure to herbicides: Baseline morbidity study results.* NTIS AD-A-138-340. Brooks AFB, TX: USAF School of Aerospace Medicine.

AFHS. 1984b. *An epidemiologic investigation of health effects in Air Force personnel following exposure to herbicides: Mortality update: 1984.* NTIS AD-A-162-687. Brooks AFB, TX: USAF School of Aerospace Medicine.

AFHS. 1991. *An epidemiologic investigation of health effects in Air Force personnel following exposure to herbicides. Mortality update: 1992.* AL-TR-1991-0132. Brooks AFB, TX: Armstrong Laboratory.

AFHS. 2005. *An epidemiologic investigation of health effects in Air Force personnel following exposure to herbicides. 2002 follow-up examination results.* AFRL-HE-BR-SR-2005-0003. Brooks AFB, TX: Epidemiologic Research Division, Armstrong Laboratory.

Batista, M. L., M. Olivan, P. S. M. Alcantara, R. Sandoval, S. B. Peres, R. X. Neves, R. Silverio, L. F. Maximiano, J. P. Otoch, and M. Seelaender. 2013. Adipose tissue-derived factors as potential biomarkers in cachectic cancer patients. *Cytokine* 61:532-539.

Battistonia, A., S. Rubattua, and M. Volpea. 2012. Circulating biomarkers with preventive, diagnostic and prognostic implications in cardiovascular diseases. *International Journal of Cardiology* 157(2):160-168.

Baylin, A., E. Kabagambe, X. Siles, and H. Campos. 2002. Adipose tissue biomarkers of fatty acid intake. *American Journal of Clinical Nutrition* 76:750-757.

Campbell, K. L., K. W. Makar, M. Kratz, K. E. Foster-Schubert, A. McTiernan, and C. Ulrich. 2009. A pilot study of sampling subcutaneous adipose tissue to examine biomarkers of cancer risk. *Cancer Prevention Research* 2:37-42.

CDC (Centers for Disease Control and Prevention). 2009. *Fourth national report on human exposure to environmental chemicals.* Atlanta, GA: CDC.

Cecil, P. F. 1986. *Herbicidal warfare: The Ranch Hand Project in Vietnam.* New York: Praeger.

Centola, M., G. Cavet, Y. Shen, S. Ramanujan, N. Knowlton, K. A. Swan, M. Turner, C. Sutton, D. R. Smith, D. J. Haney, D. Chernoff, L. K. Hesterberg, J. P. Carulli, and J. R. Curtis. 2013. Development of a multi-biomarker disease activity test for rheumatoid arthritis. *PLoS ONE* 8(4):e60635.

Chen, H., S. Werner, S. Tao, I. Zörnig, and H. Brenner. 2014. Blood autoantibodies against tumor-associated antigens as biomarkers in early detection of colorectal cancer. *Cancer Letters* 346(2):178-187.

Condorelli, G., M. V. Latronico, and E. Cavarretta. 2014. MicroRNAs in cardiovascular diseases: Current knowledge and the road ahead. *Journal of the American College of Cardiology* 63(21):2177-2187.

Cross, A. J., J. M. Major, and R. Sinha. 2011. Urinary biomarkers of meat consumption. *Cancer Epidemiology, Biomarkers and Prevention* 20:1107-1111.

Diaz, L. A., Jr., and A. Bardelli. 2014. Liquid biopsies: Genotyping circulating tumor DNA. *Journal of Clinical Oncology* 32(6):579-586.

Dobson, R. 2012. Urine: An under-studied source of biomarkers in multiple sclerosis? *Multiple Sclerosis and Related Disorders* 1:76-80.

Edwards, T., R. J. Cadigan, J. P. Evans, and G. E. Henderson. 2014. Biobanks containing clinical specimens: Defining characteristics, policies, and practices. *Clinical Biochemistry* 47:245-251.

El-Sohemy, A., A. Baylin, E. Kabagambe, A. Ascherio, D. Spiegelman, and H. Campos. 2002. Individual carotenoid concentrations in adipose tissue and plasma as biomarkers of dietary intake. *American Journal of Clinical Nutrition* 76:172-179.

Ferriby, L. L., J. S. Knutsen, M. Harris, K. M. Unice, P. Scott, P. Nony, L. C. Haws, and D. Paustenbach. 2007. Evaluation of PCDD/F and dioxin-like PCB serum concentration data from the 2001-2002 National Health and Nutrition Examination Survey of the United States population. *Journal of Exposure Science Environmental Epidemiology* 17(4):358-371.

Fortin, M. C., G. Carrier, and M. Bouchard. 2008. Concentrations versus amounts of biomarkers in urine: A comparison of approaches to assess pyrethroid exposure. *Environmental Health* 7(55):1-13.

Henderson, G. E., R. J. Cadigan, T. P. Edwards, I. Conlon, A. G. Nelson, J. P. Evans, A. M. Davis, C. Zimmer, and B. J. Weiner. 2013. Characterizing biobank organizations in the U.S.: Results from a national survey. *Genome Medicine* 5(3):1-12.

Hou, L., X. Zhang, D. Wang, and A. Baccarelli. 2012. Environmental chemical exposures and human epigenetics. *International Journal Epidemiology* 41(1):79-105.

Hsu, P. C., I. Y. Chem, C. H. Pan, K. Y. Wu, M. H. Pan, J. R. Chen, C. J. Chen, G. P. Chang-Chien, C. H. Hsu, C. S. Liu, and M. T. Wu. 2006. Sperm DNA damage correlates with polycyclic aromatic hydrocarbons biomarker in coke-oven workers. *International Archives of Occupational & Environmental Health* 79:349-356.

IOM (Institute of Medicine). 2006. *Disposition of the Air Force Health Study.* Washington, DC: The National Academies Press.

Kagedan, D., I. Lecker, C. Batruch, C. Smith, I. Kaploun, K. Lo, I. Grober, E. Diamandis, and K. Jarvi. 2012. Characterization of the seminal plasma proteome in men with prostatitis by mass spectrometry. *Clinical Proteomics* 9:2-20.

Klein, R. J., C. Zeiss, E. Y. Chew, J. Y. Tsai, R. S. Sackler, C. Haynes, A. K. Henning, J. P. San-Giovanni, S. M. Mane, S. T. Mayne, M. B. Bracken, F. L. Ferris, J. Ott, C. Barnstable, and J. Hoh. 2005. Complement factor H polymorphism in age-related macular degeneration. *Science* 308(5720):385-389.

Kruijsdijk, R. C. M., E. Van Der Wall, and F. L. J. Visseren. 2009. Obesity and cancer: The role of dysfunctional adipose tissue. *Cancer Epidemiology, Biomarkers and Prevention* 18(10):2569-2578.

Laird, P. W. 2010. Principles and challenges of genome-wide DNA methylation analysis. *Nature Reviews Genetics* 11(3):191-203.

Lavatelli, F., D. H. Perlman, B. Spencer, T. Prokaeva, M. E. McComb, R. Théberge, L. H. Connors, V. Bellotti, D. C. Seldin, G. Merlini, M. Skinner, and C. E. Costello. 2008. Amyloidogenic and associated proteins in systemic amyloidosis proteome of adipose tissue. *Molecular and Cellular Proteomics* 7(8):1570-1583.

Lavon, I., M. Refael, B. Zelikovitch, E. Shalom, and T. Siegal. 2010. Serum DNA can define tumor-specific genetic and epigenetic markers in gliomas of various grades. *Neuro-Oncology Journal* 12(2):173-180.

Mardis, E. R. 2013. Next-generation sequencing platforms. *Annual Review of Analytical Chemistry* 6:287-303.

Michalek, J. E., A. J. Rahe, and C. A. Boyle. 1998. Paternal dioxin, preterm birth, intrauterine growth retardation, and infant death. *Epidemiology* 9(2):161-167.

NCI (National Cancer Institute). 2014. *Mission and goals of the biorepositories and biospecimen research branch.* http://biospecimens.cancer.gov/about/overview.asp (accessed November 7, 2014).

Nguyen, A. H., and S. J. Sim. 2014. Nanoplasmonic biosensor: Detection and amplification of dual bio-signatures of circulating tumor DNA. *Biosensors and Bioelectronics* 67:448-449.

Novak, S., T. A. Smith, F. Paradis, L. Burwash, M. K. Dyck, G. R. Foxcroft, and W. T. Dixon. 2010. Biomarkers of in vivo fertility in sperm and seminal plasma of fertile stallions. *Theriogenology* 74:956-967.

Pavuk, M. 2006. *Testing the viability of stored frozen serum samples from the Air Force Health Study using human multi-analyte profiles (MAP).* AFRL-HE-BR-TE-2007-0015. Brooks-City-Base, TX: Air Force Research Laboratory Human Effectiveness Directorate, Information Operations and Special Program Division.

Pavuk, M., N. S. Ketchum, and K. A. Fox. 2006. Mortality in US Air Force veterans of Operation Ranch Hand. *Organohalogen Compounds* 68:2136-2139.

Pavuk, M., D. G. Patterson, Jr., W. E. Turner, L. L. Needham, and N. S. Ketchum. 2007. Polychlorinated dibenzo-p-dioxins (PCDDs), polychlorinated dibenzofurans (PCDFs), and dioxin-like polychlorinated biphenyls (PCBs) in the serum of US Air Force veterans in 2002. *Chemosphere* 68(1):62-68.

Pavuk, M., D. G. Patterson, and W. E. Turner. 2014. Serum concentrations of TCDD and other dioxin-like compounds in US Air Force veterans of Operation Ranch Hand. *Chemosphere* 102:18-23.

Perera, F., and J. Herbstman. 2011. Prenatal environmental exposures, epigenetics, and disease. *Reproductive Toxicology* 31(3):363-373.

Psychogios, N., D. D. Hau, J. Peng, A. C. Guo, R. Mandal, S. Bouatra, I. Sinelnikov, R. Krishnamurthy, R. Eisner, B. Gautam, N. Young, J. Xia, C. Knox, and D. S. Wishart. 2011. The human serum metabolome. *PLoS ONE* 6(2):e16957.

Reid, G., M. B. Kirschner, and N. Van Zandwijk. 2010. Circulating microRNAs: Association with disease and potential use as biomarkers. *Critical Reviews in Oncology/Hematology* 80(2):193-208.

RHAC (Ranch Hand Advisory Committee). 2000 (October 19–20). *Meeting minutes.* http://www.fda.gov/ohrms/dockets/ac/00/transcripts/3654t1.pdf (accessed November 7, 2014).

Robinson, J. 2007. *Overview of resources provided by AFHS to MFUA*. Document given to MFUA upon transfer of AFHS assets.

Sabetian, S., A. Ardekani, M. Hodjat, M. M. Akhondi, and H. Soltanghoraee. 2010. Comparing seminal plasma biomarkers between normospermic and azoospermic men. *Journal of Reproduction and Infertility* 11(1):39-46.

Selth, L. A., M. J. Roberts, C. W. K. Chow, V. R. Marshall, S. Doi, L. M. Vincent, L. M. Butler, M. F. Lavin, W. D. Tilley, and F. Gardner. 2014. Human seminal fluid as a source of prostate cancer specific microRNA biomarkers. *Endocrine Related Cancer* 21(4):L17-L21.

Swan, S. H., R. L. Kruse, F. Liu, D. B. Barr, E. Z. Drobnis, J. B. Redmon, C. Wang, C. Brazil, J. W. Overstreet. 2003. Study for Future Families Research Group. Semen quality in relation to biomarkers of pesticide exposure. *Environmental Health Perspectives* 111(12):1478-1484.

Tartibian, B., and B. H. Maleki. 2012. The effects of honey supplementation on seminal plasma cytokines, oxidative stress biomarkers, and antioxidants during 8 weeks of intensive cycle training. *Journal of Andrology* 33:449-461.

Urban, J. D., R. A. Budinsky, and J. C. Rowlands. 2011. Single nucleotide polymorphisms in the human aryl hydrocarbon receptor nuclear translocator (ARNT) gene. *Drug Metabolism and Pharmacokinetics* 26(6):637-645.

Wang, C., C. C. Li, W. Y. Gong, and T. Lou. 2013. New urinary biomarkers for diabetic kidney disease. *Biomarker Research* 1:9-12.

Yamazaki, K., D. McGovern, J. Ragoussis, M. Paolucci, H. Butler, D. Jewell, L. Cardon, M. Takazoe, T. Tanaka, T. Ichimori, S. Saito, A. Sekine, A. Iida, A. Takahashi, T. Tsunoda, M. Lathrop, and Y. Nakamura. 2005. Single nucleotide polymorphisms in TNFSF15 confer susceptibility to Crohn's disease. *Human Molecular Genetics* 14(22):3499-3506.

Zolas, P. T., and A. Sher. 2013. Ethyl glucuronide: A biomarker for acute alcohol ingestion. *Journal of Pharmacy and Alternative Medicine* 2(2):1-4. http://iiste.org/Journals/index.php/JPAM/article/download/6884/7023 (accessed December 30, 2014).

3

Management of the AFHS Assets Research Program

The Institute of Medicine's (IOM's) research program is a unique undertaking because rarely, if ever, has such a large transfer of both data and biospecimens (assets) and the re-creation from the ground up of such an extensive database and research program been accomplished—especially in such a short time. This chapter begins with a discussion of the process of transferring the custodianship of the Air Force Health Study (AFHS) assets to the IOM. The chapter then details the numbers of and characteristics of assets received including their condition, quality, and usability for new research. The second half of the chapter describes the research program's infrastructure, including development and implementation of the proposal and review process to allow qualified researchers access to the data and the biospecimens. The last section discusses examples of approved research studies, preliminary findings, and a summary of researchers' experience using the IOM's AFHS Assets Research Program.

TRANSFER ACTIVITIES OF THE AFHS ASSETS

Following on the findings and recommendations from the *Disposition of the Air Force Health Study* report (IOM, 2006), Congress took steps to implement its recommendations. AFHS personnel briefed the House Veterans Affairs Committee regarding their plans for close-out of the project and recommended that the Medical Follow-Up Agency (MFUA), the unit within the IOM responsible for data-intensive epidemiologic studies, be designated as the new custodian. MFUA had indicated that it would accept custodianship provided that sufficient resources accompanied it. Subsequently, Congress took steps to implement the recommendations in the report, and passed two laws regarding the AFHS data and biospecimens. The first, Public Law 109-364, section 714, directed the U.S. Air Force to transfer custodianship of the AFHS's data and biospecimens to the IOM of the National Academy of Sciences at the study's conclusion (September 30, 2007). The second law, Public Law 110-389, section 803, directed the Department of Veterans Affairs (VA) to provide funding to the IOM to maintain and manage the AFHS assets in an effort to facilitate research aimed at "understanding the determinants of health, and promoting wellness, in veterans." Appendix B contains the excerpts of those laws. The legislative wording was not restrictive regarding the type of research that could be conducted using the AFHS assets.

Several actions were initiated in early 2007 to prepare to close the project and to transfer the AFHS assets to the IOM's MFUA. Cohort members were contacted to inform them of the conclusion of the study and to request consent to transfer custody of their data and biospecimens to the IOM. The Air Force contracted with Science Applications International Corporation to build a comprehensive digital database of all their records—electronic, paper, or otherwise. A biospecimens and data transfer contract was drafted between the Air Force and the National Academy of Sciences, the IOM's parent organization, describing the general principles for the change in custodianship; it was signed in April 2007. In September 2007, a high-capacity storage drive containing the AFHS's accumulated electronic data and digitized records for cohort members who gave consent for their study assets to be transferred to the IOM was given to MFUA and placed in a secure area at the IOM until funding could be secured to begin work on the data.

Because the IOM does not have the physical means to maintain a biospecimens collection, MFUA reached an agreement and contracted with the Air Force Research Laboratory (AFRL) at Wright-Patterson Air Force Base (WPAFB) in Dayton, Ohio, to house the biospecimens. Several factors contributed to the decision to house the biospecimens at WPAFB, including willingness to invest in capital improvements that would prepare the site to hold the freezers containing the biospecimens, experience with managing biospecimens, having a secure space and a monitored environment with existing personnel and equipment, and having strong internal support for this initiative. Twenty-three ultralow temperature freezers were transported to WPAFB from Brooks Air Force Base, San Antonio, Texas, using trucks that were equipped to keep them continuously operating at –70°C.

Following the passage of Public Law 110-389, the IOM entered into negotiations with AFRL to establish a cooperative research and development agreement (CRADA) that would define the terms of AFRL's management of the biospecimens under MFUA custody. The IOM and VA entered into a separate CRADA. It took approximately 3 years for a final agreement to be reached and for VA to provide funding to conduct the IOM's work on the AFHS data and to provide AFRL with funds for the maintenance of the biospecimens. Funding for maintenance of the biospecimens and associated actions, such as aliquoting samples for researchers, at WPAFB was provided by a separate CRADA with VA.

To supplement the cost of biospecimen aliquoting and shipment, researchers are charged on a cost-recovery basis.

While the CRADAs were being negotiated, researchers had begun to contact MFUA to express interest in accessing the AFHS assets. As soon as funding was in place, MFUA was able to begin designing study procedures to make the AFHS data and biospecimens available to researchers. MFUA spent the first year developing infrastructure for the program, which included conducting an in-depth analysis of the files received, loading files containing contact information for cohort participants and building the corresponding contact database, and contacting participants to request consent for the continued use of their data and biospecimens. In addition, MFUA began loading the data text files transferred from the AFHS, which contained the data required to respond to researcher requests. An expert advisory committee was formed in late 2011 and held its first meeting in March 2012. The AFHS advisory committee consisted of eight members, including the chair. The membership changed over the years as some members rotated off and new members were appointed to replace needed expertise or include additional expertise based on program needs.

DETERMINING THE CONTENT AND VALUE OF THE TRANSFERRED AFHS ASSETS FOR RESEARCH

The AFHS data and biospecimens have long been recognized as valuable because of the availability of voluminous and detailed longitudinal data linked to numerous and unique biospecimens collected from a large sample of diverse men who were followed for more than 20 years. In addition to the contributions the AFHS assets have made to the scientific literature, it is their potential for use in research beyond the scope of the original study through the use of new assays, technologies, and methods developed since the samples were collected that continues to make the AFHS a valuable and useful resource. After the IOM became the custodian of the AFHS data and biospecimens, there were still many questions and uncertainties about the state of the data and biospecimens that had to be resolved before a research program could be realized.

AFHS Cohort Consent

As part of Public Law 109-364, the Air Force was required to contact all surviving AFHS participants to obtain their written consent for both their data and biospecimens to be transferred to MFUA within the IOM. Of the 2,758 persons who participated in at least one exam cycle, 2,488—more than 90%—cohort members gave their consent (1,031 Ranch Hands and 1,457 comparisons), indicating a high level of support for this activity. A complete set of data and biospecimens from those individuals, including deceased persons, was transferred to the IOM. Participants whose information was not transferred included those who refused (5%) and those who could not be contacted after multiple attempts (5%). Two copies of the records for individuals who could not be contacted or who did not consent to the transfer exist: one was transferred to and is held by WPAFB, and one is held at the National Archives.

Once the data and biospecimens were transferred to IOM custody, additional consent was required to enable the AFHS data and biospecimens to be used for research other than studies pertaining to herbicide exposure or military health. Participants were informed that future research would likely involve a broad range of subject matter, including studies of health and well-being not directly related to military health or herbicide exposure. The IOM consent form contained the language below, which has implications for future custodianship (see Chapter 4):

> Using funding provided by the U.S. Department of Veterans Affairs under a congressional mandate, the IOM intends to make the Health Records and Biospecimens available to qualified researchers, some of whom will not be affiliated with the National Academy of Sciences (NAS) or the U.S. Government. This Consent Form is to request your consent to the NAS to make the Health Records and Biospecimens available to researchers approved by an advisory committee created by the NAS and subject to the approval of the NAS's Institutional Review Board (IRB).

Participants were asked to indicate whether they did or did not consent to allow their health records data and biospecimens to be made available for future research. The participants were informed that an advisory committee would approve all research and that their data and biospecimens would be subject to strict standards of security. Additionally, participants were asked whether or not they would agree to be contacted for any research involving new data collection. The entire consent process took approximately 18 months to complete. For participants who died between 2007 and when the IOM completed its own consent process for the cohort, consent was not legally required, but obtained from surviving spouses if possible, and participants' data and biospecimens are available for use in research. Of the 2,488 participants who consented to have their records and biospecimens transferred to IOM custody, 2,210 (89%) gave consent to have their study materials used in future research (922 Ranch Hands and 1,288 comparisons). Moreover, 1,509 participants (61%; 606 Ranch Hands and 903 comparisons) indicated that they were willing to be contacted for research requiring new data collection.

Data Quality and Usability

Questions regarding the actual quality and usability of the data and biospecimens needed to be addressed before research using the AFHS assets could be realized. Although the Committee on the Disposition of the Air Force Health Study (hereafter referred to as the "Disposition Committee") had concluded that the medical records and other study data appeared to have been properly maintained, it was not known whether the data were in a format that would allow transfer from the system that the Air Force had used to systems used and maintained by the IOM. It was also unclear whether the data could be processed, understood, and made into a usable format by a different custodian, such as the IOM.

The Disposition Committee recommended that the AFHS electronic data be organized and fully documented, including the creation of detailed and comprehensive file specifications for data files and for each variable for each examination cycle; a codebook that would constitute a comprehensive distillation of data file contents; and a document describing the contents, format, and location of those materials that have been scanned into image files. None of those activities had been completed before termination of the AFHS in 2006, nor during the time between termination and when the contract between VA and the IOM was finalized in 2011.[1]

The documentation provided by the Air Force custodians was limited to electronic file layouts, hard copies and PDFs of the questionnaires and translation codes (not the detailed and fully documented data dictionaries and codebooks that were recommended). Because the data were transferred as an assemblage of more than 123,000 disconnected text and other related files not in database form, the first year of the program was dedicated to documenting and reading in the disparate text files. For the next 3 years, in addition to continuing to read in the text files, much time and effort was spent on file integration, data cleaning, analyses required to create researcher data requests, and quality checks. Although the AFHS data are in electronic form, the data are not standardized between cycles or in a database format. Translation of these data into such a format is challenging and, since receiving the data from the Air Force, the IOM has committed a substantial amount of effort into quality control and database creation. One large difference between the IOM's research program and other longitudinal epidemiologic studies, such as the Framingham Heart Study, is that data from the latter were collected in more organized fashion and immediately stored in study databases rather than according to the less systematic approach used in collecting the AFHS data. Beginning in the second year of the program (2012), the advisory committee began reviewing research proposals and with the first proposal approval, the program marked another milestone: the ability to process the data, compile a unique dataset for each approved study, and extract the data in a usable format for investigators.

The lack of well-documented data dictionaries led to data requests from investigators that were often vague, generic, or incomplete, referring to categories of desired information rather than specific variables. This resulted in delays in identifying the most appropriate variables for provision and subsequent requests for additional data. Some researchers provided no listing of any variables that they needed, but instead relied on the program staff to determine a listing based upon the information in the researcher's proposal. The proposal process was up-

---

[1] The IOM did not have funds to support work on the assets until the contract with VA was completed.

dated after the first few instances of nonspecific variable requests to require that researchers familiarize themselves with the available information from exams and questionnaires and, at a minimum, list the types of information and for which cycles it is requested (for example, all cancer diagnoses for cycles 1, 4, and 6). Although some investigators specified the information they needed, strict inclusion criteria at times resulted in sample sizes that were too small to be used for the research proposed. In these cases, the IOM program staff worked with the researchers, under guidance of the advisory committee, to refine the selection criteria for the study. Furthermore, some researchers requested all information available with the intent of "mining" the data for associations. The committee acknowledges that there are benefits to exploratory analyses and that they are an important part of scientific inquiry to further develop and refine hypotheses, study design, and understand the underlying data structure, but it requires that such analyses be supported by specific aims and an analysis plan. As such, the advisory committee required prospective investigators to submit hypothesis-driven proposals.

Characteristics intrinsic to the data made its distribution to researchers difficult. For example, some variables regarding basic information, such as the marital status of a participant at any given time, were simply not available because of the way the data was collected. Participants were asked their current marital status in only cycles 1 and 6. For other cycles, current marital status can only be determined by reviewing marriages and other relationships that took place after the previous cycle. Many of these seemingly basic variables could not be made available until the raw AFHS data were cleaned and processed.

Before data could be provided to researchers, each variable had to be defined along with all codes used for it. As part of ensuring accuracy and completeness for data disseminated to researchers, each file was individually reviewed. As part of the quality assurance procedures, researchers were provided with data dictionaries specific to their dataset that indicated the exact location of the item in the data file, along with the size and type of the variable. A complete dataset for an individual researcher might require integrating data from more than 30 of the original text files. Each data file generated for a research study is unique. The process of generating the data file was time consuming because each step required ad-hoc custom programming, verification, and research to ensure accurate cohort selection and researcher-specific data retrieval and analysis. For each data request, a complete Data Use Agreement had to be crafted, detailing all of the data provided, time period in which the data could be used, and policies for returning derivative files and disposing of the original data file at the study's conclusion. Data files had to be structured to respond to the varying needs and expertise of the researchers, and documentation related to the process of creating the data files was continuously updated, verified, and maintained. Although this method of distributing the data to approved investigators is one of several that could be used, the IOM program staff and advisory committee considered it to be the best approach available at present as it allows for a customizable approach to each study. The studies generally have few requested data elements in common, and this distribution method is integral to processing the biospecimen requests.

For studies that requested biospecimens, the quantity and availability of the biospecimens determined the available study population that could be selected. Integrating the physical biospecimen requirements with restrictive cohort selection criteria resulted in many additional challenges. The IOM program staff worked with the researchers to run multiple iterations of cohort selection using different parameters to finalize and select the cohort of interest. Major revisions to the study design or the cohort selection criteria had to first be approved by the advisory committee, and re-reviewed by the IRBs if applicable.

Biospecimen Quality and Usability

The Disposition Committee had also recommended that the biospecimens be inventoried and fully documented prior to the AFHS's scheduled termination, but this had not been completed when the IOM assumed custodianship of the assets. The Air Force had a record (a Microsoft Excel file) of each freezer's contents by shelf, box, and participant. However, they did not verify this information before transferring the biospecimens to AFRL. The record did not detail the biospecimen history (when it was received, whether additional aliquots had been made, number of freeze–thaw cycles, if it had been given to other investigators, and the like). The IOM created a biospecimen tracking database from the Excel file received. The database maintains location information for each participant's biospecimens, and location information is updated and maintained as requests are made and fulfilled. The database was created with a standardized volume per biospecimen type for each participant. The IOM program staff carefully review each selected participant's biospecimen availability to ensure that no one participant's biospecimens will be depleted before allowing processing by AFRL. As biospecimens continue to be selected, aliquoted, and distributed, the advisory committee has instituted preliminary cutoff volumes to ensure samples are not depleted. The IOM biospecimen database includes information for each consented individual participant; their biospecimens by cycle, type, and amount; and number of aliquots, freeze–thaw cycles, aliquot date, and to whom samples were provided. However, because the original biospecimen data file was not based upon a full inventory, the IOM biospecimen database includes errors that are not identified until a particular sample is requested. Although this strategy of updating the biospecimen database works as an interim solution, the committee concurs with previous committees that an updated biospecimen inventory is critical to maximize the efficiency of the repository and better inform future research with the knowledge of what is actually available.

The IOM program staff use the database to select the biospecimen samples that are to be processed and shipped to researchers requesting them. Multiple documents, including chain of custody, aliquot vial labels, sample request information, and protocol, are sent to AFRL that identify the participant, location of the participant's biospecimen(s) of interest, and the requested aliquot size. Printed labels are provided for all newly aliquoted samples along with chain of custody forms. One occasional problem encountered regarding requests for biospecimen samples was when the information on their location in the transferred spreadsheet did not match the actual location. For example, some biospecimens that were selected for research were missing completely, some had less volume than recorded, and some were not in the location indicated by the information sent with the biospecimens. To resolve these inherent problems and to avoid unnecessary handling, program staff identify and select additional biospecimens in the database to be potential replacements for primary selected biospecimens that cannot be located. Part of the processing procedure includes a preliminary check of requested study biospecimens by AFRL staff to ensure all of the requested biospecimens are available in the volume required before beginning the aliquoting process. Notably, the majority of biospecimens have been located and were found to be properly maintained. Those additional practices instituted by the advisory committee contributed to greater accuracy in preparing a study's biospecimens because the study cohort did not have to be reselected or changed if biospecimens were missing. It also ensured that the biospecimens were not wasted or subject to unnecessary handling. In all of the biospecimens selected for use in approved research studies—2,270—only 20 could not be found and had to be replaced.

Time Effect of Storage

The Air Force performed a reassay on selected AFHS biospecimens that had been frozen for 9–24 years to evaluate their stability, condition, and viability before the biospecimens were transferred to the IOM's custodianship (Fox and Pavuk, 2006; Pavuk, 2006). Five participants (two Ranch Hands and three comparisons) were randomly selected based on the 988 participants who met the following inclusion criteria: participated in all six physical examinations, had multiple 10 mL frozen serum specimens collected at each of the first five physical examinations, no history of cancer, had at least one quantifiable dioxin measurement, and signed an informed consent form. One 10 mL serum sample per examination per participant was selected from the first five exam cycles for a total of 25 serum samples. Stored samples were analyzed using a microsphere-based multiplexed immunoassay method to identify biomarker and protein expression patterns. Each serum specimen was analyzed for 177 analytes (78 specific serum antigens, 43 autoimmune serologies, and 56 infectious disease serologies). Sixteen analytes were measured in replicate in at least two exam cycles and were included for comparison. Eight analytes (alpha–1 antitrypsin, C3 complement, creatine kinase, IgA, IgM, PSA, SGOT/AST, and TSH) had continuous measured levels with normal, laboratory set ranges, and eight other analytes (hepatic panel, mitochondrial antibodies, and thyroidal microsomal antibodies) had a positive or negative finding in the AFHS data. Results from earlier analyses and the immunoassay panels were similar and the correlation coefficients on each of the 16 replicate analytes were statistically significant (p-value ranged from < 0.001–0.04). Of the 177 analytes examined 170 (96%) provided measurable results using the immunoassay panels. The seven analytes that did not yield measurable results are usually present only in pathologic conditions (IL-1beta, IL-2, IL-4, IL-6, MMP-9, glutathion-S-transferase, and calcitonin). Complete results were available for 147 (83%) of all analytes in analyzed samples across all five cycles. Pavuk stated that the results below standard curves appeared to be participant-related rather than related to the examination cycle or storage time, and were not related to dioxin level or status as a Ranch Hand or comparison. No difference in the number of results was observed for earlier compared with later examinations. The results suggest that the biochemical integrity of the serum samples appeared to be well preserved and remains viable given that sensitive immunoassays could be successfully performed. Similarities in results between the AFHS measurements taken at the time of collection and these later reanalyses of the same measurements were further evidence of the stability of the stored frozen serum samples.

Although the AFHS samples are relatively old (from 32 years for cycle 1 to 12 years for cycle 6), they have been stored at –70°C or below since their collection and have been found to have retained their usefulness for epidemiological studies. While many biospecimen repositories are now stored in liquid nitrogen or ultralow freezers because the colder temperatures slow the deterioration of biomarkers during long-term storage, older repositories were not traditionally stored in such low temperatures, which is another unique and valuable feature of the AFHS. Many biomarkers are expected to remain viable over time and not be adversely affected despite the age of the samples. For example, the AFHS samples from the earliest cycle have been used for genetic studies and showed that the samples were not adversely affected by long-term storage.

Quality of Biospecimens

Although the small studies of serum samples by Pavuk demonstrated that the biospecimens were viable, there was no guarantee that the rest of the serum samples or samples from the other types of biospecimens were viable. Because an updated inventory of the biospecimens, which should have included a visual inspection, was not completed prior to the transfer to the IOM, the quality of the biospecimens was unknown. The Disposition Committee visited the AFHS biospecimen repository at Brooks Air Force Base, where the biospecimens were maintained before they were moved to WPAFB under the aegis of the IOM. The committee randomly selected three samples from three unique participants in order to observe the retrieval process and inspect the preservation of the samples. The committee observed no signs of leakage or major thaws with any of the three biospecimens, but did note that the position of the fluid in one urine sample was consistent with either minor thawing and refreezing or that the initial freezing had occurred in two phases (IOM, 2006). Because only three samples were inspected out of a possible 91,500, many unknowns remained regarding the quality of the biospecimens, which could have implications for their use in new research.

Fortunately, the AFHS biospecimens had been appropriately stored and preserved and the AFHS biorepository is considered to be of comparable quality to other large repositories that house biospecimens for epidemiologic studies (IOM, 2006). To ensure the quality of the biospecimens was maintained, the IOM staff and the AFRL staff created operating procedures and protocols for aliquoting or sectioning each type of biospecimen, developed material transfer and custody agreements, and created forms to document and catalog the number of vials, volume, and location of each biospecimen that was selected for use in a research study as the database inventory was developed.

Recent and distinct studies that used the biospecimens under the IOM's research program have found that the storage conditions used over the more than 30 years have been adequate and that the biospecimens have retained a similar quality as "fresher" samples (Pavuk, 2006; Janice Chambers, Mississippi State University, personal communication, April 2014). Preliminary results from approved studies have demonstrated that adequate quantities and quality of genetic material can be successfully obtained from various biospecimens, including semen, adipose tissue, and whole blood (Kim Boekelheide, Brown University, personal communication, April 2014). These and other approved studies have confirmed that numerous biomarkers of interest are still viable, reliably measurable, and consistent in the AFHS samples. One such example is the enzyme paraoxonase 1 that was shown to have measurable, although lower, activity (18%) in serum collected in cycle 1 compared with activity measured in serum collected in cycle 6 (53%) (Chambers, 2014). Several of the assays used in approved research projects have been multianalyte assays, which provide test results for multiple analytes simultaneously on each individual sample, making the sample usage relatively efficient.

Approved research studies have used the AFHS biospecimens for epidemiologic studies of biochemical factors that were not possible at the time the biospecimens were originally collected. Biochemical measurement technologies have improved dramatically over the past 30 years, and the AFHS samples could potentially be used to assess a broad spectrum of outcomes beyond their original intent. Moreover, a broad array of other health information is available for each participant making it possible to explore new avenues of correlates to overall health.

SUMMARY OF THE APPLICATION PROCESS TO ACCESS THE AFHS ASSETS

The advisory committee was charged with fostering the use of the AFHS data and biospecimens in new research and reviewing research proposals. To fulfill its charge, it developed criteria to be used to evaluate the scientific merit of research proposals to make decisions about which would be given access. Additionally the advisory committee created an application process. Its first step was to develop and disseminate a request for proposals (RFP). The RFP was first announced on April 30, 2012. Researchers from the United States and abroad, from academic institutions, industry, and government organizations, were eligible to apply for access to the AFHS assets. The RFP was publicized on the AFHS project page of the IOM website and initially sent to researchers and organizations that were familiar with the AFHS or had an interest in research related to herbicide exposure.

Researchers were asked to first complete a letter of intent limited to 500 words that was reviewed by the committee to ensure that the primary hypothesis, research aims, and general proposed methods were well justified and scientifically sound, and that information being sought was available. Each letter of intent was assigned to two members of the advisory committee to review and evaluate independently. The reviewers for each letter of intent were specifically chosen based on their expertise that most closely matched the proposed investigation. Reviews were compiled, discussed by the entire committee, and a consensus decision was made about whether to invite the investigators to submit a full proposal. Those researchers who were approved were then invited to complete a full proposal. After experimenting with submission deadlines, beginning with the May 2013 RFP announcement, the committee chose to use a rolling submission process for both letters of intent and invited proposals. That served to both increase the number of interested researchers because they were not restricted by set deadlines, and to prevent a backlog of datasets and biospecimens to be prepared as would occur when several proposals were approved concurrently.

The same two committee members who were responsible for the primary review of the letter of intent were assigned to be the primary reviewers of the full proposal. Reviewers used the evaluation criteria outlined in the RFP, which are reproduced in Box 3-1. If necessary, a third committee member with relevant expertise was also asked to review specific sections of the proposal. For example, the committee member with biological sample repository and laboratory expertise would evaluate the parts of the proposal related to biospecimen usage. All evaluations were compiled, and the scientific merit of the proposal discussed by the full advisory committee before a decision on access to the requested assets was made. Although all proposals had to meet a minimum threshold to be approved, those proposals that requested the use of biospecimens had to meet a higher standard than proposals that requested data only because the biospecimens are a limited resource. Applications requesting biospecimens that have been judged scientifically meritorious, undergo review for feasibility testing of the requested biospecimens for the planned laboratory assays or evaluations. This feasibility assessment is conducted by committee members who are knowledgeable about the requested biospecimen types and the assays that are intended to be used. Factors considered in the review include adequacy and appropriateness of the requested

BOX 3-1
Proposal Review Criteria

For All Proposals:

1. The question being addressed is of considerable scientific and/or medical interest.
2. The researchers are familiar with the relevant literature.
3. The study design is appropriate to address the question.
4. The proposal provides evidence that the sample size is adequate for the proposed research.
5. The research team has the appropriate qualifications and experience to conduct the study.
6. The proposed work cannot be undertaken without data of the type collected as a part of the AFHS.
7. The researchers have provided sufficient evidence that they can perform to accepted standards including quality control, data security, laboratory analytical standards, and data analysis.
8. If the study requires the IOM to contact members of the cohort to obtain additional information, extra expenses may be incurred. The costs of obtaining these data and addressing any relevant ethical and consent issues are addressed adequately.
9. The researchers or their collaborators are qualified to appropriately select and perform the statistical analyses required for the study, and to interpret the results properly.
10. The researchers have identified suitably qualified and trained persons to perform the work, and agree that only these persons will have access to the data, which are either provided or generated from the biospecimens.
11. The researchers have or are able to obtain sufficient time and resources to conduct the study.

biospecimen type and amount, anticipated stability of the analyte under the known storage conditions, and known technical performance of the planned assays on the requested samples. Whenever possible, the amount of residual sample is minimized by careful planning of biospecimen amounts provided.

Proposals that included requests for pilot funds were critically reviewed for both scientific merit and to ensure that awarded pilot funds were likely to facilitate generation of preliminary data or results that could then be used to apply for larger grants to further the research, instead of solely being a mechanism to support research projects without future planning. Committee members would

If the proposed research requires access to biospecimens:

12. The researchers have provided assurance that they will maintain biospecimens under secure and confidential conditions.
13. The amounts of biospecimens requested by the researchers are appropriate for the specified study, and are not excessive, given the limited availability of the biospecimens.
14. The researchers demonstrate the feasibility of the assays or other tests proposed on the type of biospecimens available, including sensitivity, specificity, and reproducibility.
15. Methodology is detailed with clearly worded descriptions of all processes and tests to be carried out.

If the researchers are requesting pilot funding:

16. The researchers demonstrate how the pilot funding will enable them to prepare a compelling proposal for subsequent research using the AFHS data and/or biospecimens. For example, pilot funding could be used to analyze data to prepare power calculations or demonstrate that assays or other laboratory techniques would work with the AFHS biospecimens. The researchers have provided information about funding agencies that might be interested in this topic and the level/nature of funding that would be requested.
17. The researchers have detailed how the pilot funds will be used.
18. If access to data is required for sample size and power calculations, the approach is described in detail.

Overall Score: 1–9 (based on National Institutes of Health Scoring System, where 1 is the highest score and 9 is the lowest score) (NIH, 2014).

sometimes request additional information or detail from the investigators before making a final decision.

Researchers whose proposals were approved were granted access to the requested electronic data and, if applicable, biospecimens, contingent on IRB approval from both the researcher's home institution and the National Academy of Sciences' IRB. Those investigators were required to give annual updates to the advisory committee that included their accomplishments, any difficulties encountered and their resolution, and preliminary results. Communication between investigators and the committee was maintained by the IOM. Thus, the advisory committee provided scientific and administrative oversight of the research and evaluated results from both feasibility/pilot studies and larger studies.

Since the IOM became the custodian of the AFHS assets, more than 80 inquiries from researchers have been received. Many of these were related to the specific diseases and conditions that developed in the cohort over the 20-year follow-up of the AFHS, several of which had been attributed to Agent Orange exposure as outlined in IOM reports (see *Veterans and Agent Orange* series reports from 1994–2012). As of February 1, 2015, 26 letters of intent were submitted by investigators, 24 of whom were invited to submit full proposals. The Web-based system used to submit the letters of intent and proposals also tracks the number of applications started but not completed or submitted; 16 additional letters of intent were started but not submitted. Of the 24 proposal invitations, 19 complete proposals were submitted to the committee for review. Among those investigators who chose not to submit a full proposal, staff learned through communications that for at least 4 of them the lack of funding support influenced their decision, especially after pilot funds were no longer available. Of the 19 proposals reviewed by the committee, 13 were approved; 6 studies required access to the data only, and 7 studies used both data and biospecimens. Table 3-1 provides a summary of approved studies. The abstracts of approved studies are found in Appendix C. The proposals that were not approved did not meet the standards for scientific merit outlined in Box 3-1. In its decision letters to principal investigators of proposals that were not approved when first reviewed, the committee detailed its major concerns and allowed the investigators to respond and clarify any of the points and refine issues of their proposal before the committee made a final decision on access to the AFHS assets. The committee requested additional information on 11 proposals, and ultimately approved 5 of them for access to the AFHS assets.

Generating Interest

Following the announcement of the first RFP, the committee designed a two-page flyer to advertise the availability of the AFHS assets. The flyer was intended to increase awareness of the availability of the AFHS assets for researchers, particularly in fields outside herbicide exposure and toxicology. Both committee members and the IOM staff brought copies of the flyer to national meetings including the American Public Health Association, the American Geriatric Society, the American Society of Clinical Oncologists, and the American Association for Cancer Research. Project staff made presentations about the AFHS research program at national meetings to encourage use in new research directions. The committee announced the availability of the data and biospecimens for use in research on several listservs, including the VA listserv of all chiefs of staff for research, listservs targeting aging and geriatrics research, and announcements to the toxicology-research community, all of which resulted in increased interest as demonstrated by many new queries and submitted letters of intent. An invited

TABLE 3-1 Studies Approved for Use of the AFHS Data or Biospecimens Since 2012[a,b]

Principal Investigator	Institution	Proposal Title	Request Type
Batty	University of Edinburgh	Cognitive function in middle age as a predictor of later life health: Analyses of data from the Air Force Health Study	Data only
Boekelheide[c]	Brown University	Identifying epigenetic molecular markers of dioxin exposure in Vietnam veterans	Data and biospecimens
Chambers[c]	Mississippi State University	A longitudinal study of paraoxonase 1 (PON1) in relationship to type 2 diabetes and aging	Data and biospecimens
Haws	ToxStrategies, Inc.	Exposure–response relationship for dioxin and cancer and noncancer health outcomes in the Air Force Health Study cohort using physiologically based pharmacokinetic modeling of exposure and updated mortality	Data only
Knafl	University of North Carolina at Chapel Hill	Effects of dioxin exposure for male Air Force Vietnam veterans on reproductive outcomes	Data only
Mandel	Exponent, Inc.	The reanalysis of the Ranch Hand data	Data only
Mazur	Syracuse University	Testosterone changes	Data only
Mitchell	Emory University and Atlanta VA Medical Center	Identifying novel biomarkers of vulnerable coronary artery disease: The Air Force Health Study	Data and biospecimens
Ramos[c]	University of Louisville	Detection of L1 protein in Ranch Hand biospecimens	Data and biospecimens
Ross[c]	Pacific Health Research and Education Institute, VA	Parkinson's disease and pre-motor features of Parkinson's disease in the Air Force Health Study	Data only
Roth[c]	VA San Diego Healthcare System	Caveolin's role during healthy aging	Data and biospecimens
Seldin[c]	Boston University	Incidence of abnormal free light chains and other markers of light chain amyloidosis in veterans exposed to Agent Orange: A pilot study	Data and biospecimens
Shim[c]	Centers for Disease Control and Prevention	Monoclonal gammopathy of undetermined significance (MGUS) and microRNAs in Ranch Hand veterans	Data and biospecimens

[a] This list represents studies approved as of February 1, 2015.

[b] The date that a proposal was approved and the date the study was able to start varied depending on the type and extensiveness of the requested assets. For example, studies that requested more biospecimens (for example, more than 100) took longer to process than those requesting fewer samples.

[c] Denotes studies that were awarded pilot funding.

paper summarizing the accomplishments of the AFHS Assets Research Program is under review by Military Medicine.

Limited pilot funding was available in the first 3 years of the program, as was required by Public Law 110-389, section 803, although no funds were awarded until the second year. As noted above, the first year was dedicated to transferring the assets and setting up the program infrastructure. All investigators were eligible to apply for pilot funding; however, only the most meritorious proposals were awarded funds. The goal of awarding funding was to provide assistance for feasibility studies (such as for obtaining biospecimens and reagents), rather than to entirely support the proposed work. The committee could award up to $250,000 per year. Three proposals were awarded pilot funds in the second year and four proposals were awarded pilot funds in the third year. Funding ranged from approximately $12,000 to $84,000 per proposal. The availability of pilot funding served to generate additional interest in the assets from the research community.

RESEARCH APPROVED FOR THE USE OF THE AFHS ASSETS

Although the Disposition Committee recommended that a 5-year commitment should be given to creating and operationalizing the new AFHS management program, and an evaluation of its success should not occur less than 2 years after the final pilot funding awards, the VA contract supported only a 4-year commitment with a report on the feasibility and advisability of maintaining the AFHS assets for future research required in the fourth year. The first proposals were approved in 2012. Given the amount of work required to prepare the data to be used in research, this did not provide sufficient time for most researchers to complete their work. While it is premature to expect publications so soon into the program, the committee was able to review preliminary results from approved studies in the feasibility stage, and one investigator was able to publish his results (Mazur et al., 2013, 2014).

Summary of Approved Studies

The research proposals reviewed and approved for use of the AFHS biospecimens and data span a wide range of scientific and biomedical research topics (see Table 3-1 and Appendix C). The extensive data and biospecimens collected over the extended period of the AFHS makes it valuable for examining potential etiological factors relating to rare diseases or diseases and conditions that have a long latency periods, or for examining biomarkers associated with the natural history of slowly progressive disorders. For example, diagnostic data have been combined with DNA analysis from whole blood, semen, and adipose tissue samples to investigate genetic markers of disease predisposition or prognosis. Use of technology, such as next-generation sequencing, enables DNA data from an entire genome to be obtained from a whole blood sample of less than 0.5 mL

(Koboldt et al., 2013). Investigators generating whole genome data from the AFHS participants can use the subset of the data that is relevant to their study. Ideally, whole genome data generated by a single investigator will also be made available to other investigators (see Chapter 4), so that the samples do not need to be used to generate whole genome data more than once.

Just over half of the studies that the advisory committee approved were for use of the AFHS biospecimens and data, while the others requested and received data only. Several of the studies that were approved for use of the AFHS biospecimens were pilot studies for which limited preliminary data were available due to prior nonavailability of suitable biospecimens. These studies were successful in establishing the feasibility of applying novel assays to the stored biospecimens to generate interesting scientific leads. Results from pilot studies, which may be insufficient on their own to produce a research publication, provide critical information to inform next steps in these challenging areas of research.

The study of monoclonal gammopathy of undetermined significance (MGUS) is one good example. This study made use of the longitudinally collected serum samples to examine expression levels for a panel of microRNAs and other novel biomarkers for their association with development of MGUS and its progression to multiple myeloma, a disease with incidence that increases with age. Serum samples for individuals were examined at three time points to determine biomarker changes. Promising results from this pilot study could provide justification for study of these biomarkers in expanded collections of serum samples or in prospective clinical trials of MGUS.

A second approved study investigated potential etiologic factors for the rare blood disorder amyloid light chain amyloidosis, which has a reported incidence rate of 1,200 to 3,200 new cases per year in the United States (Amyloidosis Foundation, 2014). Matched sets of serum and adipose tissue specimens were required to conduct the biomarker evaluations necessary for the study. The first step was to provide 228 serum samples that had matching adipose tissue samples for the same year of collection. All serum samples were initially screened for abnormal levels of free immunoglobulin light chains, a biomarker of a plasma cell dyscrasia in the bone marrow due to amyloid light chain amyloidosis, multiple myeloma, or MGUS. Additional diagnostic tests of serum protein electrophoresis and serum immunofixation electrophoresis were performed on the six samples (3%) that had screened positive for abnormal levels of free immunoglobulin light chains, which is suggestive of the possible presence of plasma cell disease. Supplementary tests to identify biomarkers of organ dysfunction were also performed. Matched adipose tissue samples for the six serum samples that screened positive were then tested for tissue confirmation of light chain amyloidosis (David Seldin, Boston University, personal communication, April 2, 2014). The rarity of biorepositories with clinically annotated adipose tissue and matched serum samples would have rendered this research project infeasible had it not been for the AFHS resources. It would be prohibitively expensive to conduct new prospective studies to obtain the rich amount of data and unique biospecimen collections that were already available in the AFHS for these projects.

Other approved proposals examined long-term health outcomes and changes in biomarkers, DNA methylation, or other measures with regard to response or susceptibility to health endpoints (Boekelheide, Chambers, Mitchell, Ramos, Roth, Seldin, Shim); using biospecimens for early detection, identification of therapeutic targets, and/or disease response markers (Chambers, Mitchell, Seldin); and new statistical analyses, methods, and models of exposures on health effects (Haws, Mandel). Research that consisted of using new analytical techniques to examine questions that were not addressed as part of the AFHS was approved (Batty, Mazur), as was using publicly available information to expand the period of follow-up for the cohort (Haws).

The research performed using the AFHS assets in the course of the original study has benefited veterans through findings from studies of dioxin and herbicide exposure that were unique risks in this population. Several of the earliest approved studies under the IOM's program focused on the health effects and potential biomarkers of dioxin and other herbicide contaminants that may provide further evidence of association and other occupational exposures or environment effects specific to individuals' service in the Air Force during Vietnam. The fact that these were among the first proposals is not surprising given that these investigators were already familiar with the "Ranch Hand Study" and its rich data and biospecimen resources and they were specifically made aware of the RFP. For example, three research investigations proposed using different assumptions and approaches than have been applied to date to reanalyze outcomes examined by the AFHS (Haws, Knafl, Mandel). Another research study linked administrative records from VA to gather additional follow-up health and outcome information from a subset of AFHS participants (Ross). The AFHS resource remains relevant for dioxin-related research through the application of novel technologies (e.g., genome-wide epigenetic assays) to more deeply interrogate the molecular consequences of dioxin exposure, for the application of novel statistical approaches, or for further research focusing on new or longer-term disease or quality-of-life endpoints (such as reproductive outcomes).

Awareness of the AFHS resource and its potential value beyond dioxin research has been disseminated throughout the research community, as evidenced by a steady increase in the diversity and number of proposals received. The AFHS cohort is a subset of the U.S. population, and although not representative since the cohort consists of mostly white men whose average age was 64 years at the end of the original study period, it has a diversity of socioeconomic status and educational background and the members have many of the health outcomes that align with those observed in the broader similar demographic subset of the U.S. population (e.g., prevalence of heart disease and diabetes). Findings from new epidemiologic or biomarker studies continue to be relevant and applicable to veterans but are also relevant to all U.S. men of similar demographic makeup.

Additional research topics that have emerged among the submitted and approved proposals include aging, neurologic diseases, and cancer research, showing that AFHS data and biospecimens remain relevant for important biomedical and public health research projects.

Investigator Experience Using the AFHS Assets

In an effort to determine whether researchers found the AFHS to be a valuable and useful resource, the committee designed a short questionnaire and solicited feedback from the principal investigators of approved research studies after they had received all of their requested data or biospecimen samples. The committee asked these investigators to give specific examples of how the AFHS assets were unique for their research and any suggestions they could provide to make the AFHS Assets Research Program even more useful or valuable. Questions were primarily subjective and related to the overall application process, quality and completeness of the data received, quality and usefulness of the biospecimens, and future plans regarding the results and additional uses of the AFHS assets.

The overall feedback was supportive of both the AFHS assets as well as the application process. A majority of the investigators indicated that they learned about the AFHS through the IOM's project website, but a number of them were already familiar with the AFHS and learned about the opportunity for using the assets for new research through colleagues or direct inquiry. Every one of the investigators used the information found on the project website to plan their research and found it to be helpful for planning their study. All of the investigators had contacted the IOM staff before submitting a letter of intent, and none reported any issues or problems with the application, pilot funding (if applicable), or award process. Principal investigators reported that the IOM program staff was able to provide helpful guidance, and that the committee's feedback on letters of intent when making invitations to complete a full proposal was also helpful in refining project goals or improving study design or approach.

The investigators who received and used the AFHS data generally rated the completeness of it as 4 or 5 on a scale of 1–5, with 5 being very complete. While at least half of the investigators' studies were still in progress when they provided feedback to the committee, they were confident that the quality of the data, both provided and generated, would allow them to draw reliable conclusions. Even though work was continuously ongoing to prepare the data for use in research, preparing researcher data requests could take weeks to months depending on the variables of interest; however, all of the investigators who had received all of their requested materials reported that it had been provided to them in a "timely manner."

More than half of the approved research studies required the use of biospecimens in addition to data. Numbers of biospecimens requested for studies ranged from 12 to 1,163. All investigators reported that the samples were clearly labeled and in good condition and they were able to perform the proposed assays successfully on the samples. Both DNA and microRNA were isolated successfully from the samples.

The AFHS Assets Research Program began accepting applications in mid-2012, and several of the research studies are still in process (especially for those studies approved in 2013 and 2014). However, one principal investigator was able to publish two peer-reviewed articles (Mazur et al., 2013, 2014) in this relatively short time period. The other investigators have indicated that they are either in the process of drafting or submitting publication(s) or expecting to do so once their study is complete. Although no investigators have used the preliminary data generated by analyses of the AFHS assets in a subsequent grant application, it is too soon to know whether the pilot projects will produce results that lead to such applications. All but one of the investigators stated that their research could not have been completed without the AFHS research assets. The investigators, except one who was interested in pursuing a different line of research, stated that they found the AFHS data and biospecimens to be unique and valuable, and would consider using the AFHS assets for future research if possible. When asked about their overall experience with the AFHS Assets Research Program, all of the respondents characterized their experience as very good or excellent. And all of the investigators stated they would recommend use of the AFHS to others. As an example, Dr. Chambers stated, "Such cohorts and data collection are not easy to find. . . . This could be a tremendously valuable resource for answering questions . . . to the toxicology community." Dr. Chambers further stated,

> I would further urge that the samples and data base be retained for at least a few more years to allow access to this unique resource along with the determination of whether samples are still reliable. The potential information from such a longitudinal study is not available elsewhere and could yield some extremely useful scientific insights, both related to herbicide exposure (not the objective of our project) and unrelated to herbicide exposure (the objective of our project).

This chapter described the AFHS management program in terms of its ability to provide valuable data and biospecimens for research despite the many difficulties, unknowns, uncertainties, and complications that had to be addressed by the IOM. The following chapter provides an overview of possible options for the continued management of—and access to—the AFHS data and biospecimens. It also outlines several aspects of data management that need to be considered as the research program continues to grow and move forward and offers some of the many possible opportunities for future research using the AFHS data and biospecimens.

REFERENCES

Amyloidosis Foundation. 2014. *Description and background.* http://www.amyloidosis.org/treatmentinformation/primaryal.html (accessed August 5, 2014).

Chambers, J. 2014. *A longitudinal study of paraoxonase 1 (PON1) in relationship to type 2 diabetes and aging.* PowerPoint presented at the Advisory Committee meeting, Washington, DC.

Fox, K. A., and M. Pavuk. 2006. Testing the viability of stored frozen serum samples from the Air Force Health Study using human multi-analyte profiles (MAP). *Organohalogen Compounds* 68:1390-1393.

IOM (Institute of Medicine). 2006. *Disposition of the Air Force Health Study.* Washington, DC: The National Academies Press.

Koboldt, D. C., K. M. Steinberg, D. E. Larson, R. K. Wilson, and E. R. Mardis. 2013. The next-generation sequencing revolution and its impact on genomics. *Cell* 155(1):27-38.

Mazur, A., R. Westerman, and U. Mueller. 2013. Is rising obesity causing a secular (age-independent) decline in testosterone among American men? *PLoS ONE* 8(10):e76178.

Mazur, A., R. Westerman, A. Werdecker, and U. Mueller. 2014. Testosterone and type 2 diabetes in men. *Aging Male* 17(1):18-24.

NIH (National Institutes of Health). 2014. *How reviewers score applications.* http://www.niaid.nih.gov/researchfunding/grant/strategy/pages/5scoring.aspx (accessed December 10, 2014).

Pavuk, M. 2006. *Testing the viability of stored frozen serum samples from the Air Force Health Study using human multi-analyte profiles (MAP).* AFRL-HE-BR-TE-2007-0015. Brooks-City-Base, TX: Air Force Research Laboratory Human Effectiveness Directorate, Information Operations and Special Program Division.

4

Continued Management and Use of the AFHS Assets

This chapter addresses several issues key to enhancing the future use of the Air Force Health Study (AFHS) assets (the data and biospecimens) by the scientific community. The value of the assets for continued scientific research is discussed in depth in Chapter 3, but it is important to recognize that the availability and viability of such a resource depends on a careful balance between access by the scientific community and protection of the study participants. To that end, the committee begins the chapter with a discussion of options for the continued management and use of the AFHS data and biospecimens in research. That section is followed by an overview of considerations that any custodian or steward, including the Institute of Medicine (IOM), must take into account when making the assets available for research to the scientific community. The chapter ends with a discussion of opportunities for future research and possible metrics that can be applied to evaluate the future success of the program.

OPTIONS FOR CONTINUED MANAGEMENT AND USE OF THE AFHS ASSETS

The committee strongly believes that it is feasible and appropriate to maintain the AFHS data and biospecimens as their value and usefulness for research have been confirmed (as discussed in Chapter 3). The committee identified two options regarding the continued maintenance and management of the assets:

- Retain custody and management of the assets within the Medical Follow-Up Agency (MFUA) of the IOM; either maintaining the current manage-

ment structure—where MFUA curates the assets—or forming a partnership between the IOM and another organization to manage and distribute the data to researchers (as is presently done with the biospecimens); or
- Transfer custody of the AFHS data and biospecimens to another organization.

Retain Custody and Management Under the IOM

At the time when the disposition of the AFHS was first considered, seven options for an appropriate custodian were identified (IOM, 2006). These included both nongovernment entities—specifically the IOM's MFUA—and government research entities such as a Department of Veterans Affairs (VA) or VA-affiliated epidemiologic data management and distribution mechanism, a Department of Defense epidemiologic data management and distribution system, other federal agencies (such as the Centers for Disease Control and Prevention or the National Institutes of Health [NIH]) with expertise in epidemiologic data management and distribution mechanisms. The National Archives was also considered as a potential custodian, as was using the infrastructure that was in existence for the AFHS within the Air Force at the time, and identifying a new custodian through a competitive process (IOM, 2006). Each of these potential custodians had strengths and weaknesses that were discussed in the report, which did not name a recommended alternative.

Ultimately, Congress delegated custodianship of the AFHS assets to MFUA within the IOM. This was consonant with MFUA's long history (since 1946) of conducting epidemiologic studies of military and veterans' health; its experience maintaining data collections, including managing, analyzing, and disseminating data, protecting participants' privacy and confidentiality; and its independence, objectivity, and thorough review and oversight processes.

The *Disposition of the Air Force Health Study* report identified several characteristics of a good custodian for the AFHS assets. Those characteristics included demonstrated ability to properly house, secure, and manage both very large and complex data and the accompanying biospecimens (IOM, 2006).

Since the IOM became the custodian of the AFHS electronic data, the storage devices containing them have been housed in a secured area with limited access to only those individual IOM staff members that have been approved to work with them. The IOM has invested heavily in data cleaning and preparation to foster the use of the data in new research, and, although the process is ongoing, program staff have considerable expertise managing, analyzing, and disseminating the data and biospecimen samples. Chapter 3 provides details regarding the condition of the received data and the preparation, cleaning, quality checks, and other management activities required to make it usable for research.

Because the IOM does not have the physical means needed to maintain a biospecimens collection, MFUA reached an agreement and contracted with the Air

Force Research Laboratory (AFRL) at Wright-Patterson Air Force Base (WPAFB) in Dayton, Ohio, to store, maintain, and ship samples to approved researchers. The IOM provides specific instructions on which samples to process based on analysis performed by the IOM staff using researcher requirements. In addition, the IOM provides all documentation, including vial labels and chain of custody forms for sample dissemination, as well as updates the electronic database with biospecimen sample locations and amounts. The *Disposition of the Air Force Health Study* report recommended that any custodian meet four guidelines pertaining to the storage and maintenance of the biospecimens: have a secure facility with controlled access and redundant power supplies to house freezers that are maintained at least –70°C, establish protocols for continuous monitoring of freezer function and other quality control and assurance practices for long-term biospecimen preservation, have appropriate local and remote alarm systems in place to ensure the preservation, viability, and integrity of the biospecimens, and have staff trained to respond to freezer breakdowns, power outages, or other emergencies at any hour (IOM, 2006). The storage, monitoring, maintenance, and emergency response guidelines outlined in the *Disposition of the Air Force Health Study* report have been met by AFRL (AFRL, 2012, 2014b).

More than 40 operating procedures and documents have been developed in collaboration with AFRL that detail the policies and procedures that inform the continuous monitoring of the freezers, ensure alarms systems are working, and staff are able to respond quickly and appropriately to any potential emergencies concerning the freezers and biospecimens, regardless of the day or the time. Operating instructions have been drafted, reviewed, approved, applied, and are in alignment with national best practices for biological sample repositories as set forth by the International Society for Biological and Environmental Repositories and the National Cancer Institute (NCI) (ISBER, 2011; NCI, 2011) for personnel qualification and training, establishing a quality assurance system, preparing the biospecimens, packaging and shipping the biospecimens, safe handling of biospecimens and laboratory equipment, equipment maintenance, and security and monitoring. These processes and procedures for storing, monitoring, maintaining the biospecimens, and responding to emergencies require periodic review and updating to ensure they continue to conform to national best practices for biological sample repositories.

A quality assurance program has been instituted for the biospecimens to prevent inconsistencies in handling, provide standardization for conducting repository activities, and ensure best practices are used (AFRL, 2012). The program ensures that all repository systems function properly and according to guidelines. The continued monitoring and evaluation of processes and equipment allow for early identification of deviations from proper operations that may compromise the quality and integrity of the AFHS biospecimens. Biosafety cabinets are certified annually and are used to process and aliquot all of the AFHS biospecimens.

Laboratory staff notify the IOM staff about concerns regarding the freezers or biospecimen integrity and document any incidents in annual progress reports.

Standard operating procedures for thawing, freezing, and aliquoting of frozen biospecimen samples were developed and followed during the fulfillment of investigator requests (AFRL, 2013, 2014a). Investigators review the procedures that are proposed to be used on their samples in advance and sign their acknowledgement, or they propose a revised protocol that is agreed upon and then used to process the request. The first time a serum or urine biospecimen is needed from a previously unaccessed vial, it is aliquoted from its (5 mL or more) container into smaller 0.5 mL aliquots; only the amount needed is sent to the investigator, and the remainder is returned to storage in smaller aliquots. This process serves to minimize freeze–thaw cycles for subsequent requests, and thus, helps to maintain biospecimen integrity as well as facilitating biospecimen selection for future requests. The aliquoted samples are placed into storage containers that are stable for long-term storage. To date, the most commonly requested amount of serum is 0.5 mL, so larger vials are realiquoted to 0.5 mL samples. The serum sample volumes requested and used to date have left most of the biospecimen remaining for future studies. When samples of adipose tissue and semen are requested, the amount requested is removed and the rest of it refrozen without separation into additional aliquots. The committee finds the storage, monitoring, and emergency response guidelines recommended by the *Disposition of the Air Force Health Study* report have been met by AFRL and has no concerns about the maintenance of the biospecimens.

Although MFUA has much experience in storing, managing, and disseminating epidemiologic data, the committee observes that partnering with an agency or organization that administers databases or funds research as its primary mission may be in a better position to actively promote and leverage use of the AFHS data and biospecimens for new and original research than the IOM alone. This is not meant to cast doubt on the ability of the IOM to manage this program, but instead to suggest a possible alternative approach that uses the strengths of the IOM in scientific review while enhancing the breadth of the research program. In such a partnership, the IOM would retain custodianship of the AFHS assets and approve the type and amount of information given to researchers through an advisory committee. The partner organization would physically hold the electronic data; have responsibility for cleaning and applying quality control, managing, and maintaining them; preparing and distributing data (and selecting any biospecimens associated with research studies) to researchers as directed by the IOM; and apply its expertise in promoting the use of large datasets for research. Such a partnership would be similar to the one currently in place for the storage and maintenance of the biospecimens between the IOM and AFRL.

There are various examples of organizations partnering, including public and private alliances, to perform data and biorepository management. One such example is the Framingham Heart Study, a large ongoing prospective epidemiologic study that collected and stored biospecimens that can be used by outside researchers, which is maintained under the joint stewardship of Boston University and the National Heart, Lung, and Blood Institute (NHLBI) at NIH (Framingham Heart Study, 2014). The molecular genetic data, including genomic sequencing results, and gene expression and microRNA profiles collected from the Framingham Heart Study are available through NIH's database of Genotypes and Phenotypes, and exam data are available for use in research through the NHLBI. Therefore, although much of the Framingham Heart Study data and biospecimens are maintained and made available by other entities, the stewards continue to maintain overall control of the assets, and such an arrangement may be possible for the AFHS data and biospecimens. Certain organizations have experience in the management of broader data repositories, such as the Inter-university Consortium for Political and Social Research (ICPSR), housed at the University of Michigan, which holds, curates, and provides access to hundreds of diverse datasets from more than 740 universities, government agencies, and other institutions that are used in new research studies (ICPSR, 2014).

Shifting storage, curation, and maintenance of the data to another party would require the completion of the data dictionaries and other documentation activities that MFUA has been pursuing and could not take place without funding for the staff time needed to prepare the assets.

Transfer Custody

Congress may decide to designate an alternate custodian to manage the AFHS data and biospecimens and to administer the research program. Such a decision should be carefully considered and based on thorough comparison of the attributes and characteristics between the IOM and proposed alternatives. If a new agency or entity were chosen to be the custodian of the AFHS assets, there are many considerations that would have to be resolved before such a transfer could take place. First, the cohort would have to consent to have their materials transferred to a new custodian. Approximately 10% of the cohort did not consent to have their AFHS materials transferred from the Air Force to the IOM in 2007, and another 10% did not consent (or could not be located) to have their materials used in new research. These losses do not appear to have significantly affected the research designs of the types of studies proposed. However, if cohort members were asked to consent a third time, the likely result would be a further diminution of the cohort, the extent of which would be dependent on the new selected custodian. Loss of additional cohort members will reduce the value of these assets for use in additional research because the sample size may not be adequate for many types of studies. Loss of additional participants may introduce further selection bias as well as weaken the scientific validity of the analyses. Second, if the AFHS assets were transferred to a different custodian, their availability for continued use in research would likely suffer a setback because the new custodian would have to set up a new research program or integrate the AFHS into an existing one. It is unclear in what format the data would be transferred. For example, the IOM received more than 100,000 text files that had to be read in, integrated, and cleaned. Whether the custodian would choose to create a new database using the original text files or would try to incorporate the work the IOM has done (if the systems are compatible), each option has large implications for continuing a program of research using the AFHS data and biospecimens. Third, the custodian would need to find an appropriate and secure location to house the freezers in which the biospecimens are stored, form a relationship with AFRL, or to transfer them to a different ultralow temperature storage facility. As was noted above, funding would need to be provided to support the staff needed to prepare assets for the transfer and assure that it was successfully completed.

Regardless of which custodial option is ultimately selected by Congress, a number of aspects need to be considered and accounted for to ensure continued successful use of the AFHS data and biospecimens for new and innovative research. The next section highlights some of these.

CONSIDERATIONS FOR CUSTODIANS OR STEWARDS OF THE AFHS ASSETS

The IOM's AFHS Assets Research Program has continued to develop and generate interest. There are several issues that need to be considered for managing the data and biospecimens and making them available for research. They fall into three major categories and are discussed below:

- storage (including data security), data curation, and privacy protection;
- laws, rules, and policies related to biospecimen management and dissemination; and
- sustained funding to support these activities.

Storage, Data Curation, and Privacy Protection

Data Storage

Practices and technologies to enable data security have advanced substantially since the AFHS began in the early 1980s. Data must be stored in a manner that maintains its physical and logical integrity and allows it to be easily retrievable, such as in a database file format, while protecting it from external and internal threats, such as accidental or deliberate deletions. Physical measures to achieve this include buildings with restricted access, locked offices, restricting personnel who have access, password-protected computers, and firewalls. Technical measures include restricting access to files to individuals with appropriate privileges (designated by the custodian), separating personally identifiable information from the rest of the data, and ensuring that devices storing AFHS data are not connected to the Internet. Database content must also be reviewed periodically for accuracy.

Data Curation

Data curation refers to the active management of data required to maintain it for long-term preservation and reuse. There is not a single set of nationally agreed upon criteria or standards for curation of collected data, such as that of the AFHS. However, a number of organizations including leading universities, research-driven institutions, and government agencies have developed and proposed standards and guidelines for best practices that broadly aim to maximize internal data management of a study and address making it available to the scientific community. Although there are differences across the guidelines, Box 4-1 summarizes the main elements that should be considered and applied to most datasets that are to be used for broader research. Considerations related to allowing broader access to existing scientific data include adequate preparation of data and its associated documentation, protection of human subjects, formation and use of data management plans, and appropriate attribution of the data source. An important component of any strategic plan concerning the maintenance of the AFHS data and its wide scientific use is long-term curation plans that align with best practices

BOX 4-1
Considerations for Allowing Broader Access of Datasets to the Scientific Community

1. Ensure adequate preparation of data and associated documentation
2. Protect confidentiality and privacy
3. Preserve intellectual property rights and commercial interests
4. Balance demands of long-term preservation and access
5. Develop, evaluate, and use data management plans
6. Include cost of data management in funding proposals
7. Use of data use agreements that include provisions for returning and incorporating derivative data or results
8. Ensure researcher compliance with data management plans
9. Promote public deposit of data
10. Enlist private-sector cooperation to improve access
11. Develop mechanisms for identification and attribution of data
12. Encourage data stewardship workforce development
13. Engage long-term support for repository development

SOURCE: Adapted from ICPSR, 2014.

and archival standards such as those proposed by ICPSR or the Data Documentation Initiative consortium (Data Documentation Initiative, 2015; ICPSR, 2014).

Proper data documentation, quality, and accessibility are key to enhancing and enabling use internally as well as by the broader scientific community. Because of the longitudinal nature of the AFHS and the diverse manners by which data were often collected, recorded, and documented, as well as the different statistical methods used to analyze and account for missing data and loss to follow-up, the structure of the data is complex. For example, even basic demographic variables that were collected at each visit were not necessarily formatted in the same way or asked as a question in the same way. While not an insurmountable issue, it has and continues to require substantial efforts to integrate and process data over time.

Based on the IOM's experience of working with the AFHS data and biospecimens for the past 3 years, additional issues have been identified that have affected the information available for research and timeliness of fulfilling data requests, such as the lack of standardization between examination cycles and the inability to abstract information from some types of files. To achieve a more efficient operation of the AFHS data and biospecimen resource, these issues need to be addressed. First, the data are not standardized between cycles or in a database format. As a consequence, they have not been completely subject to the types of quality control procedures necessary to ensure standardization and interpretability for future investigators. Translating the data into such a format is a large undertaking, and the IOM has already committed a substantial amount of effort into quality control and database creation.

Second, medical chart information, such as electrocardiograms and radiographs, are stored in digital media files that have not been processed owing to the costs involved in abstracting and de-identifying this information, which can only be accomplished using proprietary software. The approximately 20,000 files hold potentially important variables and information, but they are not currently in an accessible form and will require further substantial investment of money and time to become usable. Translation of the files into a usable format is the first step to making this information available; however, each file has personally identifiable information (for example name, social security numbers, medical record numbers, date of birth, etc.) that must then be redacted. Redacting personally identifiable information from these images and files will be a labor intensive process because, as the IOM staff were told when these files were transferred to the IOM's custodianship, the information can be found in multiple places within the file and is not in a standard format or place in each file. Therefore, redacting the identifiable information will require custom programming by an individual staff member to ensure that all such information has been removed and that the removal does not interfere with the image. Custom programming and the need for dedicated individual staff time for translating these thousands of files is an expensive endeavor, but until interest from investigators necessitates the need for translating some of these files, the time or expense cannot be estimated. The committee believes that these records should continue to be preserved and that dedicated funding for purchasing the software and staff time to complete the abstraction should be provided by either the funding agency or the investigators who require the information as part of their work if it is concluded that there is a demand for the information contained in them. To date, no studies have been approved that require those data.

Storage and Maintenance of Biospecimens

A complete inventory of the biospecimens has not been completed before or since the biospecimens were transferred to AFRL, and would likely be a necessary condition if the biospecimens were to be moved to another facility. Commercial biological repositories typically require an up-to-date inventory of all materials along with labeling or barcoding to align with their systems in order to accept biospecimens for storage. Thus, the approximately 91,500 biospecimens in the AFHS would likely have to be inventoried and relabeled before they could be transferred to another facility. Such an inventory is important to good management practices and an exercise that will be valuable for continued use of the biospecimens in research.

The *Disposition of the Air Force Health Study* report also called for a plan detailing biospecimen access and residual return policies, including clear documentation of the entire biospecimen access application process, the review process, decision-making criteria, biospecimen processing and shipping costs, and final disposition of biospecimens (IOM, 2006). The MFUA staff with guidance from the advisory committee has operationalized each of these items.

As recommended by NCI's Best Practices for Biorepositories and Biospecimen Resources, a legacy or contingency plan for the AFHS biospecimens is part of their successful management. This is especially important because there is currently no long-term sustained funding source for the biospecimens (or data) in place (see page 79). The conflict between currently limited and unknown additional funding and stakeholders' desire that the AFHS biorepository be permanently maintained and widely used results in an ethical dilemma that further highlights the need for a long-term strategy and legacy plan. The biospecimens have been shown to retain their biochemical properties and are valuable for many types of research studies. Should funding not be continued or a new custodian chosen, the IOM will need to consult with AFRL as they develop and implement a legacy plan for the future disposition of the biospecimen repository. The advisory committee has discussed needed components of a legacy plan as part of its responsibilities, but ultimately any transfer and legacy plan will have to be enacted by the IOM and be consistent with any prior agreements, such as the need to reconsent the cohort, laws, and institutional policies that may apply.

Privacy Protection

The extensive data collected on the AFHS participants and the linkage of that data to the biospecimens provides an extremely rich resource for research. However, while those data are being made available for a broad array of research, it is critical to maintain appropriate safeguards to protect participants' privacy and confidentiality. The safeguards in place encompass not only data already available but should extend to additional data that may be generated in the future from the biospecimens or from studies that seek to collect new information on cohort participants, such as their cause of death. Thus, balancing access to the AFHS data and biospecimens while safeguarding the privacy of the AFHS participants is essential to maintaining a successful research program. Human subjects' protection is an important consideration for all studies of human health conducted under routine research conditions, and the IOM practices regarding the AFHS assets fall within normal boundaries for protection of human subjects.

Data must be de-identified as necessary to protect participant privacy and confidentiality. Variables that could lead to the re-identification of the research participants such as date of birth, geographical location, and even diagnoses of rare diseases must be removed or binned into larger categories (e.g., the first three rather than all five digits of zip code). However, overly aggressive stripping of information that does not pose a threat to participant privacy and confidentiality risks diminishing the value and usefulness of the data. A further consideration is that the AFHS is not a randomly selected cohort but rather a specific and well-defined group, especially the Ranch Hands. De-identification processes for these data must take into account that publicly available information, such as tours of duty, military occupation, and military honors, can potentially be linked to information in the database that would allow identification of a participant. For example, in 1999 the Government Accountability Office (GAO) criticized the AFHS for its slow progress of issuing results and recommended that the data be made available to the broader community of scientists (GAO, 1999). Although the then data custodians removed all personally identifying variables (e.g., names, identification numbers, addresses, and birthdates) before posting the dataset to their website, within a few months the dataset had to be removed due to concerns about insecure data sharing practices and violations of participant privacy because the small size and special nature of the cohort led to the identification of individual participants. Therefore, any custodian must consider what other publicly available data might exist that would allow individuals with knowledge of both datasets to identify particular AFHS study participants.

The policy of the IOM advisory committee has been to limit the information released to approved researchers to only that information required for the specific research question of interest. Variables that may lead to the identification of a participant are suppressed. For example, dates of birth and death are not included with any dataset, instead only the year of birth or death is included. For studies requiring matching based on closest age, the IOM staff perform the match and include the difference of the number of days between birthdates in the dataset. Furthermore, each research study is limited to include only those participants necessary for the study. For example, the study population is limited to the specific inclusion criteria approved in the proposal. A unique dataset is generated for each approved study—there is no premade dataset that is available for all approved researchers—and for each, individual participants are assigned unique case numbers so datasets cannot be combined across studies. Only the IOM data managers have the key to the case numbers. Similarly, aliquots of requested biospecimen samples are given unique identifiers so they could be tracked independently of the parent biospecimen and adhere to confidentiality and privacy provisions. These case numbers are the same as the ones used in the corresponding data provided to the researchers. Aliquots for subsequent researchers were labeled with a unique identifier for each study. The committee finds the practices in place for maintaining privacy protection of the cohort are appropriate and should be continued.

Responsibility for maintaining privacy and confidentiality extends to investigators receiving biospecimens and data. The *Disposition of the Air Force Health Study* report identified several ways in which these responsibilities could be addressed. The review process for access to the AFHS assets and the subsequent data use and asset transfer agreement stipulates that researchers must meet and comply with several criteria to ensure the privacy and confidentiality of the individuals who have provided data and biospecimens for their use. These criteria include researchers must provide a detailed description of how they intend to use the data and biospecimens and be specific about the data variables needed; receive only de-identified data and biospecimens; agree to not attempt to identify any individuals; store the biospecimens in a secure location with limited access only by authorized personnel; store data on secure computer systems and not share it (as provided or generated from the biospecimens); and have institutional review board (IRB) approval by both the researcher's institution and the National Academy of Sciences for each approved study.

Additional restrictions on access to the AFHS data and biospecimens, such as a multistep review process, can further promote data security. Other safeguards might include varying levels of access to specific types of data and using a secure data enclave (ICPSR, 2014). However, the committee cautions that data and biospecimens are only valuable and useful if they are actually used, and investigators are far less likely to pursue research on assets that present high barriers to access.

Federal Policies Related to Management and Dissemination of the AFHS Assets

Practices governing research on biospecimens and associated data in the United States are generally related to the protection of study participants and their health information and regulations for protection of human subjects in research that have been adopted by federal departments and agencies that conduct and sponsor such research (such as VA). Laws and policies related to management and dissemination of the AFHS data and biospecimens must be taken into account by the custodian or steward if the assets are to continue to be made available for research. The IOM and AFRL staff working with the AFHS assets are trained in and adhere to regulations and protocols for ensuring the privacy of the AFHS cohort participants and are in compliance with all relevant legal requirements and ethical and social norms and expectations.

Legal requirements include federal ones, such as the Policy for the Protection of Human Subjects (Common Rule), as well as state- or jurisdiction-specific laws and regulations where the data and/or biospecimens are held. Privacy standards, such as the Freedom of Information Act and the Public Health Service Act, provide additional guidance regarding the conduct of research using the AFHS assets. Although the AFHS data and biospecimens were collected by the Air Force and at that time subject to military rules and regulations related to biomedical and behavioral research, because custodianship and responsibility for these assets was transferred to the IOM, these regulations no longer apply.

The Common Rule governs most federally funded research conducted on human beings and aims to ensure that the rights of human subjects are protected during the course of a research project (IOM, 2012). Simply, it seeks to protect human subjects through requiring informed consent of the research participant and requiring review of proposed research by an IRB. Data are considered personally identifiable if the identity of a participant is or may be readily ascertained by the investigator or associated with the information accessed by the researcher (45 CFR § 46.102(f); OHRP, 2008). However, the Common Rule exempts from its requirements research that involves solely the collection or study of existing data, documents, records, pathological biospecimens, or diagnostic biospecimens, if these sources are publicly available or if the information is recorded by the investigator in such a manner that participants cannot be identified, directly or through identifiers linked to the participants (45 CFR § 46.101(b)(4)).

The Common Rule and related regulations may be applicable to future research conducted with the AFHS data or biospecimens, but it does not apply to the assets themselves. The Office for Human Research Protections of the Department of Health and Human Services has provided guidance stating that the use of biospecimens and data in research that were originally collected for different research—such as the case with the AFHS assets—is not considered research on human subjects under the Common Rule if the investigators cannot readily identify the source individuals (OHRP, 2008). Therefore, as long as personally identifiable information is removed from the biospecimen samples and associated data before they are provided to researchers, such that the individuals cannot be readily identified by the researchers, research conducted using these biospecimens or data would not constitute human subjects research under the Common Rule. To the best of the IOM's and the committee's knowledge, there have been no breaches of confidentiality concerning the data or biospecimens while the assets have been under the IOM stewardship, and the IOM has taken active steps to ensure this remains the case.

In February 2013 the U.S. Office of Science and Technology Policy (OSTP) published a memo outlining federal policy with regard to data access (Holdren, 2013). Specifically, OSTP directed federal agencies with more than $100 million in research and development expenditures to develop plans to make the published results of federally funded research freely available to the public within 1 year of publication. At the same time OSTP has asked such agencies to require researchers to account for and manage the digital data resulting from federally funded scientific research. NIH has also promulgated guidelines and policy statements that are similar to OSTP and is increasingly focused on the broad sharing of data from NIH investigations. Investigators were encouraged to use the AFHS resource, and the associated seed funding, to generate pilot data and findings for future investigations with the expectation that such investigations would be funded by various not-for-profit and governmental agencies, such as NIH.

Data-sharing expectations with respect to NIH date back to February 2003 when NIH published the final data sharing policy, which stated that studies provided with at least $500,000 in any year of a study would need to provide a data-sharing plan (NIH, 2003). In this policy, NIH stated that shared data should be devoid of identifiers that would permit linkages to individual research participants, as well as variables that could lead to "deductive disclosure" of an individual participant's identity. If data sharing could not be performed then investigators would need to provide an explanation. The IOM's AFHS Assets Research Program has already implemented measures that align with the NIH policy to prevent identification of any individual (see Privacy Protection section on page 74) including generating a unique dataset for each approved study and assigning randomly generated case numbers for each individual included in the dataset so datasets cannot be combined across studies. While grant panel reviewers have been instructed not to factor such plans into the determination of scientific merit or a priority score for a proposal, NIH program staff is responsible for overseeing the data-sharing policy and for assessing the appropriateness and adequacy of the proposed data-sharing plan.

In August 2014, NIH published a data-sharing policy that focused on genomic sequence data, which extended and revised a 2007 policy that focused mainly on genomewide scans of single nucleotide polymorphic regions (NIH, 2007, 2014). This policy states that NIH expects sharing of large-scale human genome sequencing data for broad reuse regardless of the quantity of funding provided by the institute. These policies reinforced the statement that the data should be devoid of identifiers, but went on to state "expect(s) investigators generating genomic data to seek consent from participants for future research uses and the broadest possible sharing" and that "the breadth of the sharing permitted by the consent may be taken into consideration during program priority review."

The language of the consent form used by the IOM complies with NIH and other federal policy. Consent was obtained from participants for their biospecimens and medical information to be transferred from the Air Force to the IOM. Following the transfer of the AFHS materials, the IOM embarked on an 18-month effort to obtain consent from all living participants to allow their data and biospecimens to be used for research (see Chapter 3). The language of the consent was broad as to not limit the types of research that could use the AFHS data and biospecimens, and meets the suggested guidance of the Department of Health and Human Services that has been codified in federal regulations (HHS, 2011) and the criteria suggested by NIH and other institutions to allow the distribution of properly protected data to qualified scientists. Participants were assured that all research would have to be approved by an expert advisory committee appointed by the National Academy of Sciences and subject to the approval of its IRB. If a different custodian were to be selected in the future, a new consent of the current AFHS cohort would be required. Experience has shown that with each new consent process, approximately 10% of participants do not give consent, thus, any new consent requirements would likely result in a diminution of the current cohort thereby reducing its value for continued research.

This most recent NIH policy does not require consent for data generated from de-identified clinical biospecimens and cell lines collected before the effective date of the policy, which is January 25, 2015, "because of the practical and ethical limitations in recontacting participants to obtain new consent for existing collections and the fact that such data may have already been widely used in research." From this language, it appears that NIH funding may be used to generate or study existing sequences of DNA from the AFHS biospecimens. One possibility for researchers wishing to conduct genomic studies using the biospecimens is to create a data-sharing plan that includes returning generated data to the IOM (as is currently required in the data use agreement). This would allow the generated data to be linked to the other available data which would then be made available to other researchers and prevent exhausting the biospecimens by using them to generate the same data multiple times. The data use agreement specifies that investigator-derived data does not need to be returned to MFUA until 3 years from the end of the project to allow adequate time for research results to be published. Because the AFHS Assets Research Program has only been distributing data to investigators for just over 2 years, no derivative data from researchers has been returned to MFUA or incorporated with the AFHS data files. Data use agreements and assets transfer agreements (for those studies using biospecimens), such as those currently in place between the IOM and researchers, will require periodic review and revision, to include requirements for data sharing.

Ethical and social norms and expectations include respect for the individuals who have voluntarily provided biospecimen samples, meeting public expectations of the treatment of participants and the use of their materials in research, and supporting efficient functioning of the biorepository and transparent practices.

Obtaining informed consent from the AFHS cohort members to use their data and biospecimens in research studies is one example of the ethical principle of respect for individuals and promoting their autonomy. Social norms include concepts such as engendering public trust, by being aware of and responsive to interested stakeholder groups. For example, the IOM staff have presented updates of the program and approved research to the Ranch Hand Vietnam Association (a veteran organization composed of former participants in Operation Ranch Hand) at their annual meetings to update them on the types of research questions that have been approved for use of the AFHS data and biospecimens. Other considerations include retention of participants and how biospecimens and associated data are maintained and distributed (Caulfield et al., 2014).

Sustained Funding

To properly plan for the future management of the AFHS assets, it is important to understand the limitations and conditions that were extant at the time the IOM began its stewardship of the AFHS data and biospecimens and that are likely to persist with any steward. While funding was considered an issue to the disposition of the AFHS, and rough estimates of funding levels were provided in the 2006 report, experience of the current research program has shown that the extent of work required to prepare the dataset for access was underestimated. The IOM requested approximately $4 million in total costs over 4 years to support the development of the research management program, but was awarded approximately $3 million to support the work over 4 years. Any unobligated funds remaining at the end of each contract year were deobligated or not carried forward. Because of the condition of the data received, the majority of staff effort went into cleaning and organizing the data at a level of time and effort that was not originally anticipated. Of the total award of $3.1 million, $750,000 was dedicated to funding pilot studies, which left just over $2 million over 4 years in direct costs. (This does not include costs for maintaining the biological samples, which were a separate contract between VA and AFRL.) Despite the limited resources to support it, the committee finds that the IOM has done a commendable job overcoming the problems identified before and after transfer of the AFHS assets to continue to maintain them in accordance to best practices and make them available to qualified researchers. However, sustainable funding at the level necessary to continue to manage the assets as well as make them available for researchers is necessary to continue the program.

The current AFHS research program should be viewed as a pilot effort. To truly realize the potential of the AFHS assets as a research asset like the Framingham Heart Study, the Atherosclerosis Risk in Communities Study, and other longitudinal epidemiologic studies, will require a more sustained and possibly greater investment of money and personnel. A well-defined process to access the assets is in place, but it could be improved by requiring investigators to provide data management and sharing plans, which would align with the new practices of NIH (see the Federal Policies section above). Continued resources are needed to transform the data from a disparate system of tables into a high-quality database that would facilitate rapid response to investigator requests.

FUTURE DIRECTIONS

A number of steps have been taken to prepare the data and biospecimens for use in research since the IOM began its AFHS Assets Research Program. As the research program moves forward, short- and long-term metrics are needed to evaluate its reach, value, and success. Interest in the assets is evidenced by the numbers of contacts and inquiries about them, the numbers of submitted letters of intent and ultimately approved proposals, and the amount of data disseminated. However, any evaluations of the program should consider not only the number of requests but the quality or importance of the research published from use of the assets. Since the program infrastructure has been developed and the application process is operational, the value of the AFHS research program should continue to be evaluated through the use of short- and long-term metrics. Examples of these metrics may include sustained interest in the AFHS assets quantified as described above, numbers of peer-reviewed publications, the journals that publish the results, and numbers and types (for example, R01 applications versus internal institution funding) of successful grant applications that used results from the AFHS.

Different metrics may be applied to the data and biospecimens. For example, NCI, which has funded the development and operation of many biospecimen resources, proposed guidelines for their evaluation (NCI, 2014). The questions asked of NCI-funded biospecimen resources may be adapted and used to evaluate the AFHS biospecimens:

- How effectively has the resource performed?
- What impact has the resource had on research?
- Is there a continuing need for the resource?

Effectiveness can be measured by number and variety of biospecimens or datasets provided, number of researchers served, the number of difficult to obtain biospecimens or data provided, affordability, rate of repeat requests, number of research publications generated, and usability of the provided data and biospecimens. Metrics for impact of the scientific studies conducted using the resource data and biospecimens include importance of published studies, prestige of journals, and rate of citation; role of the study in furthering useful research techniques or technologies; and whether the study led to a regulatory approval or commercialization of a product (e.g., an in vitro diagnostic or a drug), or to a new public health recommendation. Evaluation of continued need should consider whether the resource continues to see a steady flow of requests for biospecimens and data, whether the resource continues to contribute to the scientific value of studies, whether the data and biospecimen types remain relevant for contemporary scientific needs, and whether alternative sources for similar data or biospecimens are now available without prohibitive restrictions or costs. The numbers of requests and submitted applications are one measure of distribution or popularity, but are not indicative of the broader picture of value and use in the scientific community.

Opportunities for Continued Research

The longitudinal nature of the AFHS—with its extended follow-up, high rates of retention, and repeat biological samples—provides a valuable opportunity for research beyond the original aims of the study. As the AFHS research program moves forward, it can draw upon the experiences of other data resources and biorepositories to ensure success, such as the Armed Forces Institute of Pathology's tissue repository, the Atherosclerosis Risk in Communities Study, the Cardiovascular Heart Study, and the Framingham Heart Study.

There is widespread recognition that clinical and survey data collected for one study can be repurposed to study a wide array of phenotypes (Pathak et al., 2013). Moreover, there is evidence that such data can be used to support biomolecular (e.g., genomic and proteomic) association studies (Jensen et al., 2012; Kohane, 2011). Phenotypes have been shown to have merit across a wide range of topics, ranging from variations of arthritis to metabolic disorders and psychiatric issues. A certain portion of the phenotypes for which investigators provide the specifications focus on longitudinal aspects of an individual's life and there is potential that the long arc of following many of the participants in the AFHS could be studied using such methods, and has been the focus of a few of the approved proposals. Additionally, recent evidence has shown that the reuse of existing biological sample repositories and medical records can enable research studies at a significantly faster and more cost-effective pace than development of a new biological sample repository designed to target a specific disease or population (Bowton et al., 2014).

For example, the AFHS assets could be used to study the determinants of incidence or natural history of common conditions (e.g., hypertension, diabetes, and obesity) as well as correlates of health in aging. Proposals addressing these kinds of topics have been few and have only been offered recently (e.g., Chambers' on serum paraoxanase 1 in relation to type 2 diabetes and aging, and Roth's on caveolin in relation to aging) and results from those studies (other than feasibility data) are not yet available. However, the time period for evaluating the use of the data and biospecimens for other purposes has been short, and owing to the level and length of funding for this effort, this program may be more appropriately considered as a pilot research program, and could not have the impact of more established research programs.

At the same time, there are certain challenges that the AFHS research program will need to surmount in order to achieve its goals of facilitating meaningful biomedical research across a range of topics that extend far beyond investigations associated with the original purpose of data collection. An IOM report on the future uses of the Armed Forces Institute of Pathology's tissue repository covered issues that pertain to the AFHS biospecimen repository including use of the biospecimens in research while continuing to protect and respect the individuals who provided the biospecimens, meeting public expectations, and supporting the efficient functioning of the repository (IOM, 2012). Additionally, as was established in Chapter 2, many of the biospecimens in the AFHS repository are unique in type and quality. Those cases that have longitudinal data and biospecimen collection or which have concurrent matched samples of different types (e.g., blood and adipose tissue) are relatively rare and use of such valuable cases may warrant additional considerations. A highly visible marketing strategy will likely serve to increase interest in and use of the AFHS assets for new research. The next chapter summarizes the committee's findings, conclusions, and recommendations regarding the value of the AFHS data and biospecimens and their continued maintenance and use for research.

REFERENCES

AFRL (Air Force Research Laboratory). 2012. *Air Force Health Study Quality Assurance Program.* Wright-Patterson AFB, OH: AFRL.

AFRL. 2013. *Processing of AFHS specimen requests.* Wright-Patterson AFB, OH: AFRL.

AFRL. 2014a. *Processing of AFHS adipose tissue specimen requests.* Wright-Patterson AFB, OH: AFRL.

AFRL. 2014b. *AFHS Annual progress report.* Wright-Patterson AFB, OH: AFRL.

Bowton, E., J. R. Field, S. Wang, J. S. Schildcrout, S. L. Van Driest, J. T. Delaney, J. Cowan, P. Weeke, J. D. Mosley, Q. S. Wells, J. H. Karnes, C. Shaffer, J. F. Peterson, J. C. Denny, D. M. Roden, and J. M. Pulley. 2014. Biobanks and electronic medical records: Enabling cost-effective research. *Science Translational Medicine* 6(234):234.

Caulfield, T., S. Burningham, Y. Joly, Z. Master, M. Shabani, P. Borry, A. Becker, M. Burgess, K. Calder, C. Critchley, K. Edwards, S. M. Fullerton, H. Gottweis, R. Hyde-Lay, J. Illes, R. Isasi, K. Kato, J. Kaye, B. Knoppers, J. Lynch, A. McGuire, E. Meslin, D. Nicol, K. O'Doherty, U. Ogbogu, M. Otlowski, D. Pullman, N. Ries, C. Scott, M. Sears, H. Wallace, and M. H. Zawati. 2014. A review of the key issues associated with the commercialization of biobanks. *Journal of Law and Biosciences* 1(1):94-110.

Data Documentation Initiative. 2014. *Best practices.* http://www.ddialliance.org/getting-started#best (accessed January 16, 2015).

Framingham Heart Study. 2014. *Participating investigators.* https://www.framinghamheartstudy.org/about-fhs/participating-institutions.php (accessed November 4, 2014).

GAO (General Accounting Office). 1999. *Agent Orange: Actions needed to improve communications of Air Force Ranch Hand Study data and results.* GAO/NSIAD-00-31. Washington, DC: GAO.

HHS (Department of Health and Human Services). 2011. Human subjects research protections: Enhancing protections for research subjects and reducing burden, delay, and ambiguity for investigators. *Federal Register* 76(143):44512-44531.

Holdren, J. P. 2013. *Memorandum: Increasing access to the results of federally funded scientific research February 22, 2013.* Office of Science Technology and Policy. http://www.whitehouse.gov/sites/default/files/microsites/ostp/ostp_public_access_memo_2013.pdf (accessed December 1, 2014).

ICPSR (Inter-university Consortium for Political and Social Research). 2014. *Elements of a public access plan for scientific data*. ICSPR Data Management and Curation. http://www.icpsr.umich.edu/icpsrweb/content/datamanagement/ostp.html (accessed September 22, 2014).

IOM (Institute of Medicine). 2006. *Disposition of the Air Force Health Study*. Washington, DC: The National Academies Press.

IOM. 2012. *Future uses of the Department of Defense Joint Pathology Center Biorepository*. Washington, DC: The National Academies Press.

ISBER (International Society for Biological and Environmental Repositories). 2011. *Best practices for repositories: Collection, storage, retrieval and distribution of biological materials for research*. http://c.ymcdn.com/sites/www.isber.org/resource/resmgr/Files/ISBER_Best_Practices_3rd_Edi.pdf (accessed December 1, 2014).

Jensen, P. B., L. J. Jensen, and S. Brunak. 2012. Mining electronic health records: Towards better research applications and clinical care. *Nature Review Genetics* 13:395-405.

Kohane, I. S. 2011. Using electronic health records to drive discovery in disease genomics. *Nature Review Genetics* 12:417-428.

NCI (National Cancer Institute). 2011. *NCI best practices for biospecimen resources*. http://biospecimens.cancer.gov/bestpractices/2011-NCIbestpractices.pdf (accessed December 1, 2014).

NCI. 2014. *How the NCI evaluates NCI-supported resources*. http://www.cancerdiagnosis.nci.nih.gov/humanSpecimens/establish/evaluates.htm (accessed July 31, 2014).

NIH (National Institutes of Health). 2003. *Final NIH statement on sharing research data*. NOT-OD-03-032. Bethesda, MD: NIH.

NIH. 2007. *Policy for sharing data obtained in NIH supported or conducted genome-wide association studies (GWAS)*. NOT-OD-07-088. Bethesda, MD: NIH.

NIH. 2014. *NIH genomic data sharing policy*. NOT-OD-14-124. Bethesda, MD: NIH.

OHRP (Office for Human Research Protections). 2008. *Guidance on engagement of institutions in human subjects research*. http://www.hhs.gov/ohrp/policy/engage08.html (accessed November 20, 2014).

Pathak, J., A. N. Kho, and J. C. Denny. 2013. Electronic health records-driven phenotyping: Challenges, recent advances, and perspectives. *Journal of the American Medical Informatics Association* 20:e206-e211.

5

Findings, Conclusions, and Recommendations

This chapter builds on the 2006 *Disposition of the Air Force Health Study* report and the foundation laid in Chapters 1–4 to provide findings, conclusions, and recommendations in response to Congress's directive of "assessing the feasibility and advisability of conducting additional research on the assets transferred to the Agency from the Air Force Health Study (AFHS)."

OVERVIEW OF THE COMMITTEE'S APPROACH AND FINDINGS

To carry out its charge, the committee responsible for this report drew upon their professional background, training, and expertise, along with the experience garnered by the advisory committee that facilitated and oversaw research on the assets over the previous 3 years. Its findings fall into two overarching categories:

1. the scientific value of the AFHS data and biospecimens, and
2. the lessons learned in managing access to the assets.

Specifics of these are delineated below.

Scientific Value

The committee responsible for the 2006 *Disposition of the Air Force Health Study* (hereafter, *Disposition*) report faced several uncertainties when they offered their observations regarding future research prospects for the AFHS data and biospecimens. The available evidence allowed them to conclude that the assets had been well maintained and had been successfully used by the research team that gathered and analyzed them. It was less clear, however, whether the study participants would be willing to allow their materials to be used once the original AFHS concluded, whether the data and biospecimens collected could be successfully managed by another custodian, and whether the research community would be interested in and able to use them in future studies. These uncertainties have now been largely resolved.

The committee finds:

> A great majority of AFHS cohort members continue to allow use of their data and biospecimens for research.

As Chapter 3 notes, approximately 80%[1] of cohort members consented to both have their data and biospecimens transferred to the Institute of Medicine's (IOM's) custodianship and allow for these materials to be used in future research studies. The relatively high degree of continued participation improves the potential use of the assets because a larger study population allows more confident estimation of its statistical characteristics.

> The biochemical integrity of the AFHS biospecimens appears to be well preserved, and the biospecimens are amenable to analysis by long-established and newly developed assays.

A 2006 small-scale analysis (Fox and Pavuk, 2006; Pavuk, 2006) found that the AFHS serum samples stored 9–24 years provided detectable results in a high-density immunoassay, suggesting that it might be possible to use them in present-day studies. Subsequent work by investigators who were granted access to these assets under the IOM research program has shown this is still the case, with assays performed successfully with high levels of sensitivity in multiple pilot studies. Investigators were able to successfully isolate genetic material from multiple tissues and use it for their research. Because the volumes of biospecimens needed for such work are modest related to the total volume in the biorepository, the AFHS collection should be usable for several years to come.

> A broad spectrum of the scientific community has demonstrated interest in performing research on the AFHS data and biospecimens.

The intent of the original Air Force Health Study was to determine whether exposure to the herbicides used during the Vietnam War was associated with adverse health outcomes. The *Disposition* report committee concluded, though, that its potential was much greater. It speculated that a broad range of researchers—

[1] 2,210 of the 2,758 persons who participated in at least one exam cycle.

including academic, government, private sector, and foreign investigators—might be interested in performing a variety of studies using the AFHS assets, including

- reanalysis of outcomes examined by the AFHS using different assumptions and approaches than have been applied to date,
- new analyses of the medical records and other study data that examine questions that were not addressed in the AFHS,
- new studies of the collected biospecimens that take advantage of advances in technology and science to conduct analyses that were not contemplated in the AFHS protocol, and
- expansion of the study's period of analysis through follow-up of the cohort using publicly available information (IOM, 2006, p. 6).[2]

All of these possibilities have been realized in the time since the IOM has made the assets available.

The *Disposition* report did not speculate on the number and type of studies that might be proposed for and conducted with the assets, something that would in turn influence the amount of staff time required to support those studies. It instead simply noted that the amount of support required to fund a data management and access operation would depend on the demand for the assets. The fact that the number of studies that have been proposed to date is relatively small has been a mixed blessing. While a higher demand would have been evidence of greater interest among researchers it is unclear whether it would have been possible to meet it, given the initial difficulties in rendering the data into a form amenable to analysis by investigators other than the original study team. The committee believes that the breadth of the interest demonstrated over the past 2.5 years in both research topics and researcher institutions (academic, government, and industry) is indicative of untapped potential but it remains to be seen whether higher visibility in the research community will result in greater demand. Expanding the number of users of the AFHS assets will be necessary to more fully evaluate the usefulness and value of these materials for new and cutting-edge research, and thereby determine whether the cost and effort of maintaining the data and biospecimens is justified.

[2] The report also suggested that additional follow-up of health outcomes in the AFHS participants was possible, but a study of this type has not yet been proposed. If such a study were proposed and approved, it would be limited to only those participants who gave their consent to be recontacted and the questions or information requested would have to be approved by the advisory committee. The IOM staff would contact the participants and remove any personally identifiable information before providing it to the investigators.

Assets Management

The lessons learned by the advisory committee and the IOM staff involved in managing the AFHS assets will facilitate their future dissemination. These are summarized in the findings below.

The committee finds:

> The IOM staff have begun the process of transforming and standardizing the AFHS data into a form that is more amenable to modern data curation and research techniques.

The AFHS began in the early 1980s and used the technology and practices that were available to store information. While some of this was migrated to more modern forms, the data transferred to the IOM were in a format that was not amenable to current-day data management and analysis software, and were not documented in a way that facilitated their ready use. As a result, considerable time and effort had to be devoted to transforming this material into a form that could be disseminated to investigators.

The current committee feels strongly that if the organization and documentation activities for both the electronic data and biospecimens had been implemented as was recommended in the *Disposition* report (IOM, 2006, pp. 85–87), the IOM's research program would have been better equipped to begin enabling research on these assets sooner, and would have had fewer data quality problems to resolve. The IOM data staff has focused its efforts on data processing and integration, which has been a complex and time-consuming process. Although the current practice of managing the electronic data and responding to investigator requests is not an ideal model in terms of efficiency, it was the most appropriate model in terms of staff time and the condition of the data and resulted in successful completion of investigator data requests. As the number of data requests was relatively small, the current model for distributing data to investigators was effective, but reevaluating this model may be necessary as the program moves forward and the number of investigators increases.

However, additional value could be derived from the assets by rendering some records—for example, personal medical records provided in hand-coded form by participants and images of X-rays stored in a proprietary software format—into forms that would be usable by investigators, if demand for the information justified the investment.

> The IOM's data management practices have preserved the security of the AFHS data and biospecimens and the privacy of the participants' information.

The *Disposition* report's review of the Medical Follow-Up Agency (MFUA) as a candidate custodian of the AFHS assets noted that it is "the nongovernmental body with the longest history of veterans' health research [and] has proven track records of protecting participants' privacy and confidentiality" (IOM, 2006, p. 157). This is the result of a number of steps taken to protect such assets. The National Academy of Sciences' institutional review board (IRB) oversees and approves all data uses. The AFHS records with personally identifiable information are stored on a stand-alone hard drive that is accessed through a computer that has no connection to the internet and that is located in a room secured with a cyberlock. Files containing personally identifiable information are coded and are located separately from those containing other information. The biospecimens are maintained in a secured space in a building that is within a military base. Data are only released to investigators in de-identified form, and the variables provided are limited to those absolutely necessary to perform the proposed work. The identifiers used to distinguish participants' data and biospecimens are different for every investigator and study to minimize the ability to combine information. Data are transmitted in coded form and can only be decoded with a separately provided key. Investigators are subject to data use agreements—and material use agreements if they are provided with biospecimens—that govern their use of the assets. Provisions include prohibitions against transferring the data to other investigators or using it for other purposes, and requirements to return or destroy the data at the completion of the research project. Taken together, these practices help ensure the security of these assets and the privacy of cohort members' information. To the committee's and the IOM's knowledge, there have been no data breaches while the assets have been under the custodianship of MFUA.

> The Air Force Research Laboratory's management of the AFHS biospecimens has been successful.

The serum, whole blood, urine, semen, and adipose tissue specimens collected in the course of the AFHS have been maintained by the Air Force Research Laboratory (AFRL) at Wright-Patterson Air Force Base since they were transported there at the conclusion of the original study. While AFRL personnel are responsible for the day-to-day upkeep of the biospecimens, they are under the administrative control of MFUA and can only be disseminated at MFUA's direction. A cooperative research and development agreement governs the working relationship between the parties and includes training requirements for AFRL staff and requirements for periodic reports that catalog all accessions of the samples and any difficulties with the ultralow temperature freezers that hold them. Public Law 110-389 includes a provision that dedicates funding for their upkeep that is provided by the Department of Veterans Affairs (VA). The mandate for this funding, like that which supports the committee's operation and MFUA's assets management effort, ends on June 30, 2015.

The storage, monitoring, and emergency response guidelines recommended by the *Disposition* report (IOM, 2006, pp. 134–135) have been implemented by AFRL. The biospecimen repository has been well maintained and has not experienced any equipment failures or any loss of biospecimens. AFRL staff have worked with MFUA staff to establish and implement biospecimen dissemination procedures and have provided timely distribution of biospecimens. Some problems have been encountered in locating specific samples in freezers owing to inaccuracies in the manifest provided by the original study. While these are corrected when found, the committee observes that a prospective, thorough inventory of biospecimens would allow quicker access and more accurate sample selection for studies.

> Pilot funding, provided by VA under congressional direction, catalyzed research involving biospecimens. However, it is premature to draw any conclusions about the effectiveness of the pilot funding effort.

The committee responsible for the *Disposition* report expressed the belief that pilot funding was "needed to establish the AFHS data assets as a resource for independent researchers" and that it would "stimulate prospective researchers to also seek external funding from other existing sources for further or more in-depth projects" (IOM, 2006, p. 168). It recommended that Congress allocate a minimum of $250,000 per year of direct costs for 3 years for small grants for secondary data analysis or pilot projects using the data assets of the AFHS, a recommendation that Congress adopted as part of Public Law 110-389.

As noted above and discussed in Chapter 3, considerable effort was required to render the AFHS data into a form that was amenable to distribution and analysis. This, combined with the other work needed to build the research program from scratch, prevented any data, biospecimens, or funds from being disseminated in the first year of the program. The remaining 2 years attracted interest from researchers: six of the seven studies that were approved for use of data and biospecimens in 2012–2014 were awarded pilot funding; one data-only study also obtained support. This and the feedback received from investigators suggest that pilot funding was an influential factor in stimulating research on the AFHS assets.

While the committee believes that the support offered and provided to some investigators did inspire interest in the AFHS assets, the sample size (seven funded studies) is too small to draw firm conclusions on this issue. It is also too soon to know whether the results of this effort will stimulate interest in larger and more in-depth research projects in the future, something that will depend at least in part on the results of the pilot studies. In any event, it will be necessary for future researchers to obtain funding from other, outside sources if they are to pursue work on the data or the biospecimens.

The IOM has created knowledge about the AFHS assets management process that will benefit future efforts.

In the 3 years that the IOM has exercised management responsibility for the AFHS assets, it has, with the guidance of the advisory committee and cooperation of AFRL,

- improved the accessibility of the data to outside researchers by transforming it into a format that is compatible with present-day software systems and built data dictionaries that more thoroughly document it;
- corrected errors in the manifest that lists the location of stored biospecimens;
- created protocols for managing the secure storage and dissemination of the AFHS data and biospecimens; and
- developed and implemented procedures for soliciting and reviewing proposals to perform research on the assets, and facilitated the work of investigators who submit successful applications.

The experience preparing and delivering assets to successful applicants has also allowed the IOM staff to form a more realistic view of the level of effort needed to perform this work. The lessons learned have been applied to refine and streamline the operation over time and exist as a resource for whoever manages them in the future.

CONCLUSIONS

Based on the findings articulated above, the committee draws four conclusions, highlighted below.

It is possible to manage the AFHS assets and perform high-quality scientific research with them.

The committee that wrote the *Disposition* report presumed that a successful AFHS management program could be implemented but had no way of knowing whether it would work as they envisioned. That presumption has now been confirmed and the once-speculative potential of the program has been largely realized. The assets have generated continuous interest from researchers, as evidence by the more than 80 inquiries and requests between 2012 and 2014, something that is especially encouraging considering that it has not been possible to actively promote their availability until recently. While few research results had been produced to date because investigators have had relatively little time to perform their work, the initial results appear promising and the studies in progress have the potential to advance medical and scientific understanding.

Sustaining access to the AFHS biospecimens and data repository benefits the veterans community and the public at large, who gain from the information derived from studies of the assets.

The research performed using the AFHS assets in the course of the original study has benefited veterans through findings from studies of dioxin and herbicide exposure that were unique risks in this population and of their service in Operation Ranch Hand. It is too soon to know whether and which studies of the assets under the IOM's AFHS Assets Research Program will yield information that will promote health and wellness in veterans or in the greater population, but the research performed to date has the potential to contribute to the body of medical knowledge, especially the results of studies that are examining potential biomarkers of disease. This is an area of research that has greatly advanced in recent years, allowing studies that were not possible when the original AFHS was conducted. The combination of longitudinally collected data and biospecimens are well suited to biomarker research, which could advance the science regarding disease diagnosis and treatment.

These conclusions and the findings that underlie them lead the committee to the following response to the question posed by the Congress in Public Law 110-389:

The AFHS assets have been underutilized, and the custodian should continue to seek ways to improve management approaches to maximize the use of this resource in research.

While the committee is pleased that a broad spectrum of investigators is carrying out a variety of studies using the AFHS assets, it believes that the full potential of the assets has not come close to being realized. Despite the fact that the AFHS is among the few long-term longitudinal prospective studies that included longitudinally collected biospecimens, anecdotal experience suggests that few outside of the veterans health community are aware of it and its availability to outside researchers. This is likely due to both the limited marketing of the assets in professional journals, conferences, and other forums frequented by the scientific community and a general perception that data derived from military populations is only applicable to those populations. Publication of the results of currently ongoing research should help to address this problem but a more concerted effort to raise the asset's visibility to the broader research community is also needed to maximize the use of the resource. Any attempts to expand the number of researchers, though, will have to be accompanied by plans for how to fulfill requests for assets and for how to fund the work.

It is feasible and advisable to maintain the AFHS data and biospecimens and make them available for continued use in research.

The congressionally directed transfer of the AFHS assets to the custodianship of the IOM has resulted in a promising research program that is yielding further benefits from the government's approximately $143 million investment in the original study. Many of the assets have now been transformed and documented in a manner that facilitates research by outside investigators, and sustaining access to them will advance knowledge regarding the determinants of health and wellness.

The committee responsible for the 2014 IOM report *Veterans and Agent Orange: Update 2012*[3] also supported the idea that continued research use of the assets was advisable, recommending that "[a]vailable information should be gleaned from existing cohort studies" (p. 942). The report went on to specifically cite the AFHS and stated that

> A strong commitment by the federal government is required to provide sufficient funds to develop the infrastructure necessary to meet the goals of further research that uses these invaluable data and biospecimens. (p. 942)

The value of the AFHS assets lie in their combination of longitudinal epidemiologic data and associated biospecimens, and would be greatly diminished if the participants' serum, whole blood, urine, semen, and adipose tissue samples are not maintained in addition to their health data and other personal information. Both resources should thus be preserved, and the funding to maintain the resources must include support for both if the assets are to remain valuable for research. A multiyear commitment to funding is highly desirable as this would permit the long-term planning and sustaining of the infrastructure needed to conduct an effective assets dissemination program. Because the AFHS followed a cohort of active duty military personnel and veterans, and because the legislation that authorized the transfer was initiated in the Senate Committee on Veterans' Affairs, it is not unreasonable that VA would be tasked with providing the initial support to maintain the assets and fund research using them. However, as the AFHS assets have also been found to be suitable for research addressing a broader variety of topics addressing the health of the general public, another entity may be equally appropriate to continue to provide all of the ongoing funding to support this resource. The National Institutes of Health (NIH), for example, is an appropriate candidate to consider because several of its component institutes and centers

[3] The *Veterans and Agent Orange* series of reports is a comprehensive review of the literature addressing exposure to the herbicides used during the Vietnam War and their contaminants and adverse health outcomes.

address topics—cancer, heart and lung disease, aging, environmental exposures, and the like—that are amenable to study with the assets.[4]

RECOMMENDATION

Based on its findings and conclusions,

The committee recommends that Congress continue to support the maintenance of the Air Force Health Study data and biospecimens as a resource for research and to facilitate making them available to the scientific community as broadly as possible.

This must be done in a manner that continues to preserve the privacy of study participants and the security of their data.

The committee identified two options regarding the continued maintenance and management of the AFHS assets:

1. Retain custody of the assets within MFUA of the IOM, either maintaining the current management structure—where MFUA curates the assets—or forming a partnership between the IOM and another organization to manage and distribute the data to researchers (as is presently done with the biospecimens); or
2. Transfer custody of the AFHS data and biospecimens to another organization.

The *Disposition* report identified several characteristics of a good custodian for the AFHS assets. Those characteristics included a demonstrated ability to properly house, secure, and manage both very large and complex data and the accompanying biospecimens (IOM, 2006). The committee finds that the IOM has met each of these criteria and has maintained and managed the AFHS assets in accordance with best practices.

MFUA has much experience in storing, administering, and disseminating epidemiologic data, and maintaining the current structure of the assets management program is a viable option. However, joining forces with an organization that administers databases or funds research as its primary mission may allow for better promotion and dissemination of the AFHS data and biospecimens for new and original research than could be accomplished by the IOM alone. This is not meant to cast doubt on the ability of the IOM to manage this program, but instead to suggest a possible alternative approach that combines the strengths of

[4] Two NIH institutes collaborated with the AFHS over the course of the original study: the National Institute of Environmental Health Sciences (Longnecker and Michalek, 2000; Michalek et al., 2001a,b) and National Institute of Dental and Craniofacial Research (Kingman et al., 2005).

the IOM in scientific review with the institutional capabilities of a dedicated data archive. In this scenario, the IOM would retain custodianship and partner with an organization that has the infrastructure for storing, curating, and maintaining the electronic data, and experience in promoting the use of large datasets. Such a partnership would be similar to the one currently in place for the storage and maintenance of the biospecimens between the IOM and AFRL. Under such an arrangement, the IOM would retain custody of the assets and approve the type and amount of information given to researchers through an advisory committee. The partnering organization would physically hold the assets; have responsibility for cleaning and applying quality control, managing, and maintaining them; and preparing and distributing data (and selecting applicable biospecimen samples) to researchers as directed by the IOM. Affiliating with a clearinghouse for data and biospecimens such as these could significantly raise the visibility of the AFHS assets and spur their greater use.

The committee does not have any recommendations concerning partnering organizations, but can offer some examples for consideration should this option be pursued. The National Institute on Aging sponsors the Chicago Core for Biomarkers in Population-Based Aging Research (NIA, 2014), which—among other activities—maintains a directory of longitudinal studies that include biomarker and personal health data that are made available to investigators. The National Heart, Lung, and Blood Institute's Biologic Specimen and Data Repositories Information Coordinating Center serves a similar function for studies that hold assets relevant to the institute's mission (NHLBI, 2014). And the Inter-university Consortium for Political and Social Research "stores, curates, and provides access to scientific data so others can reuse the data and validate research findings" (ICPSR, 2014). The organization's vast collection of databases, which are primarily focused on health (including aging) and social science issues, includes some that address health questions and a small number with associated biospecimens. All are made available to interested investigators according to the rules set out by the data depositor. A thorough evaluation of partnering options would require discussions with candidate organizations to determine their willingness and capacity to take on curation of the AFHS data and a determination of the logistics and costs of preparing them for the relocation—tasks that are beyond the scope of this committee.

If Congress chooses to designate an alternate custodian to manage the AFHS data and biospecimens and to administer a research program, such a decision should be carefully considered and based on thorough comparison of the attributes and characteristics between the IOM and the alternatives it evaluates.[5] Many issues would have to be resolved before such a transfer could take place. A new custodian would need to exercise the same commitment to protecting the confi-

[5] Chapter 6 of the *Disposition* report contains a thorough analysis of potential alternative custodians (pp. 149–158) that the committee believes is still valid.

dentiality of the assets as has been practiced to date; and to adhere to best practices, as well as legal and regulatory requirements for the storage, maintenance, and dissemination of data and biospecimens. The cohort would have to consent to having responsibility for their study materials transferred to a new custodian, which based on previous experience would likely result in a further diminution of the cohort. Loss of additional cohort members will reduce the value of these assets for use in additional research because the sample size may not be adequate for many types of studies. Loss of additional participants may introduce further selection bias, which has not been problematic to date, as well as weaken the scientific validity of the analyses. Additionally, if the AFHS assets were transferred to a different custodian, their availability for continued use in research would likely suffer a setback because the new custodian would have to set up a new research program or integrate the AFHS into an existing one. It is unclear in what format the data would be transferred and whether the custodian would choose to create a new database using the original text files or would try to incorporate the work MUFA has done (if the systems are compatible). Finally, a new custodian would need to find an appropriate and secure location to house the freezers in which the biospecimens are stored, or to transfer them to a different ultralow temperature storage facility. Funding would need to be provided to support the staff needed to prepare assets for the transfer and assure that it was successfully completed. Given the issues that a new custodian would have to navigate, the committee observes that the feasibility and advisability of maintaining the AFHS assets for research and maximizing their use for continued research may be best met if the IOM's MFUA remains custodian but partners with an agency, organization, or institution knowledgeable in and with the infrastructure to support curation of the electronic data.

No matter who serves as custodian of the AFHS assets, a multiyear commitment to funding this organization is highly desirable as this would permit the long-term planning and sustaining of the infrastructure needed to conduct an effective assets dissemination program. The committee believes that a full 5 years of sustained access is appropriate in order to allow enough time for prospective investigators to be made better aware of the assets, develop research hypotheses, propose studies, receive approval for analyzing them, obtain funding, perform work, and disseminate results through publication in peer-reviewed journals and other venues. A reassessment of the support for maintaining and disseminating the assets should be carried out after this time period to evaluate whether the level of interest in them justifies continuing the investment of time and money needed to support their maintenance.

To derive the maximum value from research conducted on the assets, steps should be taken to make sure that the knowledge developed from them is widely disseminated through peer-reviewed publications and other professional venues and integrated into the database for subsequent use by other investigators. Data use agreements with investigators should continue to require that newly derived data be returned to the custodian, with a reasonable amount of time allotted for publications to first be generated. After these data are integrated, their availability should be promoted. When investigators establish that the biospecimens are amenable to analysis by particular assays, this information should be made generally available so other potential researchers can factor it into their plans.

Finally, the committee believes it important to increase the breadth of scientific disciplines that are aware of and use the AFHS assets. These data and biospecimens have been used to make important contributions to the understanding of the toxicologic mechanisms and health impacts of exposure to herbicides and their contaminants. However, there are many additional opportunities to learn from these assets beyond the purpose of their initial collection or use technology and science that had yet to be developed when the original study ended. Given the rapid development of new analysis techniques and instrument technologies; assay methods with improved analytical sensitivity and specificity; and an abundance of existing proteomic, metabolomic, and genetic methods that can be applied to biospecimens and data from the AFHS, the potential value and utility of the AFHS biospecimens is far greater than what was originally imagined at the beginning of the study. It will be up to the custodian to identify new areas of research and to find ways to reach the people who conduct it so they can start taking advantage of this rare and valuable resource.

REFERENCES

Fox, K. A., and M. Pavuk. 2006. Testing the viability of stored frozen serum samples from the Air Force Health Study using human multi-analyte profiles (MAP). *Organohalogen Compounds* 68:1390-1393.

ICPSR (Inter-university Consortium for Political and Social Research). 2014. *ICPSR data management and curation*. https://www.icpsr.umich.edu/icpsrweb/content/datamanagement/index.html (accessed November 24, 2014).

IOM (Institute of Medicine). 2006. *Disposition of the Air Force Health Study*. Washington, DC: The National Academies Press.

IOM. 2014. *Veterans and Agent Orange: Update 2012*. Washington, DC: The National Academies Press.

Kingman, A., J. W. Albers, J. C. Arezzo, D. H. Garabrant, and J. E. Michalek. 2005. Amalgam exposure and neurological function. *Neurotoxicology* 26(2):241-255.

Longnecker, M. P., and J. E. Michalek. 2000. Serum dioxin level in relation to diabetes mellitus among Air Force veterans with background levels of exposure. *Epidemiology* 11(1):44-48.

Michalek, J. E., F. Z. Akhtar, M. P. Longnecker, and J. E. Burton. 2001a. Relation of serum 2,3,7,8-tetrachlorodibenzo-*p*-dioxin (TCDD) level to hematological examination results in veterans of Operation Ranch Hand. *Archives of Environmental Health* 56(5):396-405.

Michalek, J. E., N. S. Ketchum, M. P. Longnecker. 2001b. Serum dioxin and hepatic abnormalities in veterans of Operation Ranch Hand. *Annals of Epidemiology* 11(5):304-311.

NHLBI (National Heart, Lung, and Blood Institute). 2014. *Welcome to the Biologic Specimen and Data Repository Information Coordinating Center (BioLINCC) website*. https://biolincc.nhlbi.nih.gov/home (accessed November 24, 2014).

NIA (National Institute on Aging). 2014. *Search population-based studies collecting biomarkers*. http://biomarkers.health-studies.org/studydemo.php (accessed November 24, 2014).

Pavuk, M. 2006. T*esting the viability of stored frozen serum samples from the Air Force Health Study using human multi-analyte profiles (MAP)*. AFRL-HE-BR-TE-2007-0015. Brooks-City-Base, TX: Air Force Research Laboratory Human Effectiveness Directorate, Information Operations and Special Program Division.

A

Committee Member Biographies

David J. Tollerud (*Chair*) is Professor and Chair of the Department of Environmental and Occupational Health Sciences at the University of Louisville School of Public Health and Information Sciences. He received his M.D. from Mayo Medical School in 1978 and his M.P.H. from the Harvard School of Public Health in 1990. Dr. Tollerud has extensive clinical training, with specialty board certifications in internal medicine, pulmonary and critical care medicine, and environmental and occupational medicine, with a long-standing interest in the health of disadvantaged and underserved populations. Dr. Tollerud's research interests include the effects of environmental pollution on asthma and other health problems, particularly among children and inner-city disadvantaged populations; nanoparticles; and stem-cell transplants. A second area of research interest is strategies to prevent work-related injury and illness. Dr. Tollerud has published more than 90 peer-reviewed scientific articles, books, and book chapters, is a member of numerous professional and scientific organizations, and sits on a number of local, national, and international committees dealing with environmental, occupational, and public health issues. He has chaired several Institute of Medicine (IOM) committees, including most recently the Committee on the Disposition of the Air Force Health Study, the Committee on the Management of the Air Force Health Study Data and Specimens (standing committee), the Committee on Epidemiologic Study: Shipboard Hazard and Defense II, and the Committee on the Long-Term Health Consequences of Exposure to Burn Pits in Iraq and Afghanistan.

Mary N. Haan is Professor and Vice Chair of Epidemiology in the Department of Epidemiology and Biostatistics at the University of California, San Francisco, where she has been on faculty since 2009. Before this appointment, she was a professor of epidemiology at the University of Michigan's Graduate School of Public Health, where she taught and directed a T32 interdisciplinary program on aging and public health. Her previous positions have also included faculty at the University of California, Davis, School of Medicine; an investigator at Kaiser Permanente's Division of Research; and as an epidemiologist at the California Department of Health Services. Her research career has focused on health disparities in relation to ethnicity and race in aging populations. She has been involved in several major longitudinal studies of vascular and metabolic risk factors in aging populations, including as principal investigator of the Sacramento Longitudinal Study on Aging, which she began in 1997 and is ongoing. Her areas of expertise include the epidemiology of aging, dementia, diabetes, and race and ethnic disparities. She has and continues to serve as a reviewer of Alzheimer's disease and other conditions related to aging for the National Institutes of Health, including the National Institute on Aging study sections, as well as for other national and international research agencies. Dr. Haan has more than 160 peer-reviewed publications, chapters, and books. She was President of the Society for Epidemiologic Research for 3 years and has had active leadership roles in the Alzheimer's Association, Conference Series on Aging in the Americas, and the International Stroke Association. Dr. Haan received her M.P.H. in Public Health Policy in 1975 and her Dr.P.H. in Epidemiology in 1986 from the University of California, Berkeley.

Kenneth W. Kizer is a Distinguished Professor at the University of California, Davis, School of Medicine, and the Betty Irene Moore School of Nursing, and the Director of the Institute for Population Health Improvement, UC Davis Health System. He is an internationally respected health care leader and one of very few persons elected to both the IOM of the National Academy of Sciences and the National Academy of Public Administration. Dr. Kizer's professional experience includes positions in academia, philanthropy, and the public and private sectors. His previous positions have included President, CEO, and Chairman of Medsphere Systems Corporation, a leading commercial provider of open source health information technology; founding President and CEO, National Quality Forum, a national quality improvement and consensus standards setting organization; Under Secretary for Health, Department of Veterans Affairs, and CEO of the nation's largest health care system, in which capacity he engineered what is widely considered the largest and most successful health care "turnaround" in U.S. history; Director, California Department of Health Services; and Director, California Emergency Medical Services Authority. He has served on the U.S. Preventive Services Task Force and as Chairman of The California Wellness Foundation, the nation's largest philanthropy devoted solely to health promotion and disease prevention, as well as on the governing boards of managed care and health information technology companies, several foundations and various professional associations, and nonprofit organizations. He also has served as an advisor to numerous foreign countries on health-related matters. Dr. Kizer is an honors graduate of Stanford University and the University of California, Los Angeles, the recipient of two honorary doctorates, a fellow or distinguished fellow of 10 professional societies, and is board certified in six medical specialties and/or subspecialties. He has authored more than 400 original articles, book chapters, and other reports. He has served on numerous IOM committees, including most recently the Board on the Health of Select Populations (standing committee), the Medical Follow-Up Agency Advisory Committee (standing committee), and the Committee on the Management of the Air Force Health Study Data and Specimens (standing committee); as well as the Committee on the Assessment of Ongoing Efforts in the Treatment of PTSD; the Committee on the Assessment of Readjustment Needs of Military Personnel, Veterans, and Their Families; and the Committee on the Long-term Effects of Blast Exposures.

Diane Leong is the Director of Biobanking and Human Biological Sample Management group at Gilead Sciences, Inc. In her current position, Dr. Leong is responsible for establishing business processes and infrastructure for the collection and management of biological samples from clinical trials and research collaborations to support biomarker group research. Prior to her current position, Dr. Leong worked at several pharmaceutical companies, where she served as the associate director of several laboratories and has expertise in the preparation of DNA for single nucleotide polymorphism genotyping for disease predisposition and pharmacogenetics studies, and high-throughput pharmacogenetics and pharmacogenomics analysis of clinical trial patient samples. Dr. Leong has a long history of published papers and abstracts, has been awarded four patents for the detection of bacteria in different body fluids, and has received numerous awards for her work. Dr. Leong received her B.A. in bacteriology from the University of California, Berkeley, and in 1985 received her Ph.D. in microbiology and molecular genetics from Harvard University.

Bradley Malin is an Associate Professor and the Vice Chair of Biomedical Informatics in the School of Medicine, an Associate Professor of Computer Science in the School of Engineering, and an Affiliated Faculty Member of the Center for Biomedical Ethics and Society at Vanderbilt University. He is the founder and current director of the Health Information Privacy Laboratory (HIPLab), an interdisciplinary endeavor that was established to address the growing need for data privacy research and development for the rapidly expanding health information technology sector. In addition to its role as a scientific research program, for the past several years the HIPLab has functioned as a data privacy consultation service for the Electronic Medical Records and Genomics network, a consortium sponsored by the National Human Genome Research Institute and National Institute of General Medical Sciences. Under the direction of Dr. Malin, the HIPLab has made contributions to a number of health-related areas, including intelligent auditing technologies to protect electronic medical records from misuse in the context of primary care, as well as algorithms to formally anonymize patient information disseminated for secondary research purposes. Dr. Malin was honored as a recipient of the Presidential Early Career Award for Scientists and Engineers, the highest honor bestowed by the U.S. government on outstanding scientists and engineers beginning their independent careers. Dr. Malin completed his education at Carnegie Mellon University, where he received his B.S. in biological sciences, his M.S. in data mining and knowledge discovery, a master's of philosophy in public policy and management, and his Ph.D. in computer science.

Lisa M. McShane is a Mathematical Statistician in the Biometric Research Branch, Division of Cancer Treatment and Diagnosis at the National Cancer Institute (NCI), where she works on the development of statistical methods for the evaluation of cancer biomarkers. She is also a statistical reviewer for biomarker study protocols and a statistical advisor to Cancer Diagnosis and Cancer Therapy Evaluation Programs. Prior to her current position at NCI, Dr. McShane was a Senior Staff Fellow in the Clinical and Diagnostic Trials Section of the Biometry Branch, Division of Cancer Prevention and Control, where she worked on developing statistical methods and collaborative research for cancer prevention trials and epidemiologic studies. In 1995, she was awarded with the National Institutes of Health Award of Merit, and in 2008 she received the National Institutes of Health Director's Award in recognition of work in the development and application of innovative trial designs for predictive biomarkers to make personalized medicine a reality. Dr. McShane received her B.A. in mathematics from Millersville State College and her M.S. in statistics from University of Kentucky. In 1989, she graduated with a Ph.D. in statistics from Cornell University.

F. Javier Nieto is Chair of the Department of Population Health Sciences at the University of Wisconsin (UW), where he has been since 2002. Prior to this appointment, he was on faculty in the Department of Epidemiology at the Johns Hopkins Bloomberg School of Public Health and has served as a member of the editorial board of the *American Journal of Epidemiology*. Dr. Nieto's main areas of research interest include cardiovascular disease epidemiology, markers of subclinical atherosclerosis, emerging risk factors for cardiovascular disease (homocysteine, inflammation markers, chronic infections), health consequences of sleep disorders, and psychosocial stress. He is also interested in methodological issues in survey research and epidemiology and in the teaching of epidemiologic methods. He is the director of the Survey of the Health of Wisconsin and has served on the IOM Committee on the Development of a Consensus Case Definition for Chronic Multisymptom Illness in 1990–1991 Gulf War Veterans. Dr. Nieto received his M.D. from the University of Valencia, Spain, and completed a residency in family and community medicine in Spain and an M.P.H. degree in

Havana, Cuba. He earned an M.H.S. in 1989 and a Ph.D. in epidemiology in 1991 from Johns Hopkins University.

Michael A. Stoto is a Professor of Health Systems Administration and Population Health at Georgetown University. A statistician, epidemiologist, and health policy analyst, Dr. Stoto's research includes methodological topics in epidemiology, statistics, and demography; research synthesis/meta-analysis and other analytical methods related to comparative effectiveness research; community health assessment; risk analysis and communication; and performance measurement. His research interests include public health practice, especially with regard to emergency preparedness; drug and vaccine safety; infectious disease policy; and ethical issues in research and public health practice. Dr. Stoto is the co–Principal Investigator of the Centers for Disease Control and Prevention–funded Linking Assessment and Measurement to PHEP Systems Improvement (LAMPS) center based at the Harvard School of Public Health, and the director of the LAMPS Systems Improvement project, which seeks to adopt systems for improving public health emergency preparedness. He serves as a consultant to AcademyHealth's Agency for Healthcare Research and Quality–funded Electronic Data Methods Forum, and is the oversight liaison director for Education, Training and Professional Development of the Food and Drug Administration/Georgetown Center of Excellence in Regulatory Science and Innovation (CERSI). Dr. Stoto is also an Adjunct Professor of Biostatistics at the Harvard School of Public Health and an adjunct faculty member of the Georgetown Public Policy Institute. He previously served on the faculty of Harvard's John F. Kennedy School of Government, the George Washington University School of Public Health and Health Services, and the RAND Graduate School. From 1987 to 1998 he was a professional staff member at the IOM, where he served as director of the Board on Health Promotion and Disease Prevention. Dr. Stoto is a Fellow of the American Statistical Association.

Linda D. Youngman is the Branch Chief for Community Grants and Program Development at the Center for Substance Abuse Prevention, which is a section of the Substance Abuse and Mental Health Services Administration of the Department of Health and Human Services (HHS). Previously she was a specialist in human clinical trials at the HHS Office of Research Integrity, and also served as a toxicologist and risk analyst in the Office of New Animal Drug Evaluation, Center for Veterinary Medicine at the Food and Drug Administration (FDA). At FDA, Dr. Youngman previously served as the Deputy Director, Acting Director, and Director of the Office of Research and chairperson of the National Antimicrobial Resistance Monitoring System. From 1990 to 2000, she was Director of Laboratories and Senior Research Fellow in the Clinical Trial Service and Epidemiological Studies unit at the University of Oxford, England, where she received tenure in 1995. Dr. Youngman received her Ph.D. in toxicology and her M.S. and B.S. in biochemistry from Cornell University. She also holds a certificate in epidemiology and medical statistics from the London School of Hygiene and Tropical Medicine and a certificate in risk assessment from the University of Maryland. Her research has been published in numerous journals, including *New England Journal of Medicine, Journal of the American Medical Association, Lancet, British Medical Journal, Proceedings of the National Academy of Sciences of the United States of America, Cancer Research*, and the *American Journal of Epidemiology*.

B

Excerpts of Public Laws Regarding the National Academy of Sciences and the Air Force Health Study

Public Law 108-183
Veterans Benefits Act of 2003

SEC. 602. STUDY ON DISPOSITION OF AIR FORCE HEALTH STUDY.

(a) Study Required.—The Secretary of Veterans Affairs shall, in accordance with this section, carry out a study to determine the appropriate disposition of the Air Force Health Study, an epidemiologic study of Air Force personnel who were responsible for conducting aerial spray missions of herbicides during the Vietnam era.

(b) Study Through National Academy of Sciences.—Not later than 60 days after the date of the enactment of this Act, the Secretary shall seek to enter into an agreement with the National Academy of Sciences, or another appropriate scientific organization, to carry out the study required by subsection (a).

(c) Elements.—Under the study under subsection (a), the National Academy of Sciences, or other appropriate scientific organization, shall address the following:

(1) The scientific merit of retaining and maintaining the medical records, other study data, and laboratory specimens collected in the course of the Air Force Health Study after the currently-scheduled termination date of the study in 2006.

(2) Whether or not any obstacles exist to retaining and maintaining the medical records, other study data, and laboratory specimens referred to in paragraph (1), including privacy concerns.

(3) The advisability of providing independent oversight of the medical records, other study data, and laboratory specimens referred to in paragraph (1), and of any further study of such records, data, and specimens, and, if so, the mechanism for providing such oversight.

(4) The advisability of extending the Air Force Health Study, including the potential value and relevance of extending the study, the potential cost of extending the study, and the Federal or non-Federal entity best suited to continue the study if extended.

(5) The advisability of making the laboratory specimens of the Air Force Health Study available for independent research, including the potential value and relevance of such research, and the potential cost of such research.

(d) Report.—Not later than 120 days after entering into an agreement under subsection (b), the National Academy of Sciences, or other appropriate scientific organization, shall submit to the Secretary and Congress a report on the results of the study under subsection (a). The report shall include the results of the study, including the matters addressed under subsection (c), and such other recommendations as the Academy, or other appropriate scientific organization, considers appropriate as a result of the study.

Public Law 109-364
John Warner National Defense Authorization Act for Fiscal Year 2007

SEC. 714. TRANSFER OF CUSTODY OF THE AIR FORCE HEALTH STUDY ASSETS TO MEDICAL FOLLOW-UP AGENCY.

(a) Transfer.—

(1) Notification of participants.—The Secretary of the Air Force shall notify the participants of the Air Force Health Study that the study as currently constituted is ending as of September 30, 2006. In consultation with the Medical Follow-up Agency (in this section referred to as the "Agency") of the Institute of Medicine of the National Academy of Sciences, the Secretary of the Air Force shall request the written consent of the participants to transfer their data and biological specimens

to the Agency during fiscal year 2007 and written consent for the Agency to maintain the data and specimens and make them available for additional studies.

(2) Completion of transfer.—
Custodianship of the Air Force Health Study shall be completely transferred to the Agency on or before September 30, 2007. Assets to be transferred shall include electronic data files and biological specimens of all the study participants.

(3) Copies to archives.—The Air Force shall send paper copies of all study documents to the National Archives.

(b) Report on Transfer.—

(1) Requirement.—Not later than 30 days after completion of the transfer of the assets of the Air Force Health Study under subsection (a), the Secretary of the Air Force shall submit to the Committees on Armed Services of the Senate and the House of Representatives a report on the transfer.

(2) Matters covered.—At a minimum, the report shall include information on the number of study participants whose data and biological specimens were not transferred, the efforts that were taken to contact such participants, and the reasons why the transfer of their data and specimens did not occur.

(c) Disposition of Assets Not Transferred.—The Secretary of the Air Force may not destroy any data or biological specimens not transferred under subsection (a) until the expiration of the one-year period following submission of the report under subsection (b).

(d) Funding.—

(1) Costs of transfer.—The Secretary of Defense shall make available to the Air Force $850,000 for preparation, transfer of the assets of the Air Force Health Study, and shipment of data and specimens to the Medical Follow-up Agency and the National Archives during fiscal year 2007 from amounts available from the Department of Defense for that fiscal year. The Secretary of Defense is authorized to transfer the freezers and other physical assets assigned to the Air Force Health Study to the Agency without charge.

(2) Costs of collaboration.—The Secretary of Defense may reimburse the National Academy of Sciences up to $200,000 for costs of the Medical Follow-up Agency to collaborate with the Air Force in the transfer and receipt of the assets of the Air Force Health Study to the Agency during fiscal year 2007 from

amounts available from the Department of Defense for that fiscal year.

Public Law 110-389
Veterans' Benefits Improvement Act of 2008

SEC. 803. MAINTENANCE, MANAGEMENT, AND AVAILABILITY FOR RESEARCH OF ASSETS OF AIR FORCE HEALTH STUDY.

(a) Purpose.—The purpose of this section is to ensure that the assets transferred to the Medical Follow-Up Agency from the Air Force Health Study are maintained, managed, and made available as a resource for future research for the benefit of veterans and their families, and for other humanitarian purposes.

(b) Assets From Air Force Health Study.—For purposes of this section, the assets transferred to the Medical Follow-Up Agency from the Air Force Health Study are the assets of the Air Force Health Study transferred to the Medical Follow-Up Agency under section 714 of the John Warner National Defense Authorization Act for Fiscal Year 2007 (Public Law 109-364; 120 Stat. 2290), including electronic data files and biological specimens on all participants in the study (including control subjects).

(c) Maintenance and Management of Transferred Assets.—The Medical Follow-Up Agency shall maintain and manage the assets transferred to the Agency from the Air Force Health Study.

(d) Additional Near-Term Research.—

(1) In general.—The Medical Follow-Up Agency may, during the period beginning on October 1, 2008, and ending on September 30, 2012, conduct such additional research on the assets transferred to the Agency from the Air Force Health Study as the Agency considers appropriate toward the goal of understanding the determinants of health, and promoting wellness, in veterans.

(2) Research.—In carrying out research authorized by this subsection, the Medical Follow-Up Agency may, utilizing amounts available under subsection (f)(1)(B), make grants for such pilot studies for or in connection with such research as the Agency considers appropriate.

(e) Additional Medium-Term Research.—

(1) Report.—Not later than March 31, 2012, the Medical

Follow-Up Agency shall submit to Congress a report assessing the feasibility and advisability of conducting additional research on the assets transferred to the Agency from the Air Force Health Study after September 30, 2012.

(2) Disposition of assets.—If the report required by paragraph (1) includes an assessment that the research described in that paragraph would be feasible and advisable, the Agency shall, utilizing amounts available under subsection (f)(2), make any disposition of the assets transferred to the Agency from the Air Force Health Study as the Agency considers appropriate in preparation for such research.

(f) Funding.—

(1) In general.—From amounts available for each of fiscal years 2009 through 2012 for the Department of Veterans Affairs for Medical and Prosthetic Research, amounts shall be available as follows:

(A) $1,200,000 shall be available in each such fiscal year for maintenance, management, and operation (including maintenance of biological specimens) of the assets transferred to the Medical Follow-Up Agency from the Air Force Health Study.

(B) $250,000 shall be available in each such fiscal year for the conduct of additional research authorized by subsection (d), including the funding of pilot studies authorized by paragraph (2) of that subsection.

(2) Medium-term research.—From amounts available for fiscal year 2012 for the Department of Veterans Affairs for Medical and Prosthetic Research, $200,000 shall be available for the preparation of the report required by subsection (e)(1) and for the disposition, if any, of assets authorized by subsection (e)(2).

C

Abstracts of Approved Proposals

Principal Investigators: David Batty and Beverly Roberts, University of Edinburgh

Title: *Cognitive Function in Middle Age as a Predictor of Later Life Health: Analyses of the Data from the Air Force Health Study*

Assets Used: Electronic data only

Abstract: This study is looking at the general association between cognitive function in midlife and later adult health: How do social and personal factors contribute to differences in human mortality, health, and health behaviors? Specifically, the researchers are examining three hypotheses. (1) Relative to their lower-performing counterparts on standard tests of cognitive function available in the AFHS (e.g., Wechsler adult intelligence scale), men with higher scores will have more favorable levels of health behaviors that have been linked to mortality, that is, they will be less likely to smoke, and more likely to be physically active, and have a healthy diet. They will also examine the links between cognition, alcohol intake, and drug use (licit and illicit). (2) Relative to their lower-performing counterparts on tests of cognitive function, men with higher scores will have more favorable levels of physiological factors that have been linked to mortality. These physiological outcomes include lower levels of blood pressure, lipids, blood glucose, insulin, and inflammatory and coagulation markers. They will separate the effects of cognition on physiological factors from those for the health behaviors. (3) Relative to their lower-performing counterparts on tests of cognitive function, men with higher scores will have longer life expectancy, that is, a lower risk of all-cause mortality and major causes of death (such as cardiovascular disease and cancer).

Principal Investigator: Kim Boekelheide, Brown University

Title: *Identifying Epigenetic Molecular Biomarkers of Dioxin Exposure in Vietnam Veterans*

Assets Used: Electronic data and biospecimens; pilot funding awarded

Abstract: Dioxin is a tumor promoter, alters immune function, and modifies metabolism. Environmental exposures modify epigenetic programming, including DNA methylation; therefore, measuring DNA methylation marks can provide insight into the cellular response to chemical exposures. This pilot project proposes using a new epigenome-wide molecular approach to detect dioxin exposure-related alterations in DNA methylation in biospecimens from the Air Force Health Study. One potential target for dioxin-related exposure effects is the epigenome. DNA methylation is one of the main types of epigenetic modifications, a process important to gene regulation. The pattern of DNA methylation reflects both inherited imprints and environmental conditions, allowing cells to adjust their programs in response to external cues. Immune cell function, tumor suppressor gene regulation, and metabolic gene expression are all altered by dioxin exposure, and these effects are almost certainly associated with changes in DNA methylation in cells of multiple tissues. We have published numerous studies of DNA methylation marks in normal and abnormal tissues, including adipose tissue, blood cells, and sperm. These novel results and the availability of unique biospecimens from the Air Force Health Study inspired us to collaborate on this effort to study the effects of dioxin exposure on DNA methylation marks in Vietnam veterans. Past studies have examined AFHS samples of adipose tissue, blood, and semen for their cellular and phenotypic attributes, but little molecular characterization has been performed using modern high-throughput techniques. This pilot project interrogates the DNA methylation marks of adipose tissue, blood cells, and sperm from highly exposed and comparison Air Force veterans. Importantly, all three tissue samples will be taken from the same study participant, allowing for both within individual and interindividual comparisons of dioxin-related effects across tissue type and by exposure level. DNA methylation profiling will use the Illumina Infinium HumanMethylation450 BeadChip assay that interrogates over 485,000 unique DNA methylation sites and provides CpG coverage of 99% of RefSeq genes in the human genome. Biostatistical analysis will identify alterations in DNA methylation marks associated with dioxin exposure.

Principal Investigator: Janice Chambers, Mississippi State University

Title: *A Longitudinal Study of Paraoxonase 1 (PON1) in Relationship to Type 2 Diabetes and Aging*

Assets Used: Electronic data and biospecimens; pilot funding awarded

Abstract: The proposed project will investigate the serum enzyme paraoxonase 1 (PON1), a hydrolase associated with the high density lipoprotein particle. PON1 has a lipolactonase activity that allows it to be protective of the cardiovascular system and its role in cardiovascular health is well documented. PON1 is also implicated in reducing risk for type 2 diabetes. PON1 activity has been found to decline with age, but evidence for this has not resulted from a longitudinal study of the same individuals. Therefore, the AFHS offers an opportunity to investigate both the role of PON1 in risk of type 2 diabetes development and changes in PON1 activity with aging. The first hypothesis proposed for this project is: Lower serum PON1 levels are associated with the development of type 2 diabetes. The second hypothesis is: serum PON1 activities decline with age. This proposed case-control study will analyze serum spectrophotometrically for PON1 activities assayed with three substrates: paraoxon, phenyl acetate, and dihydrocoumarin. C-reactive protein (a marker of inflammation) and PON1 protein will be quantified by ELISA. The serum samples will be from males of similar age from the control group who participated in cycles 1 and 6, half of whom were diabetic in 2002. We propose a longitudinal assessment of PON1 activities and PON1 protein and C-reactive protein concentrations in the same individuals over 20 years, in relationship to occurrence of type 2 diabetes. Logistic regression models relating these measurements with clinical and demographic information will be developed. Serum and demographic/clinical information will be requested for 250 participants (125 diabetics and 125 nondiabetics). (Initially a sample activity validation study with 25 samples will be conducted to verify that activities and protein levels are sufficiently robust before the remaining 225 samples are obtained.) Outcomes anticipated from this study are (1) documentation on whether lower PON1 activity and/or protein levels are associated with occurrence of type 2 diabetes, and (2) whether PON1 activity and/or protein levels decline with age in participants without type 2 diabetes. If these associations are supported in this longitudinal study, then PON1 (with further documentation) may be a useful new biomarker for type 2 diabetes risk. We anticipate that the data and preliminary models derived from this project would be used as preliminary data to seek support for additional studies using the AFHS biospecimens and data.

Principal Investigator: Laurie Haws, ToxStrategies, Inc.

Title: *Exposure-Response Relationship for Dioxin and Cancer and Non-Cancer Health Outcomes in the Air Force Health Study Cohort Using Physiologically-Based Pharmacokinetic Modeling of Exposure and Updated Mortality*

Assets Used: Electronic data

Abstract: To date, very few studies have been published that describe the original TCDD exposures experienced by the Ranch Hand cohort based on the measures of TCDD in serum, and none have applied the current state-of-the-art physiologically-based pharmacokinetic (PBPK) models to estimate internal TCDD exposure over time. This exposure metric can be used with health outcome data to quantify the risk of cancer and other health outcomes, such as type 2 diabetes, in the cohort. Furthermore, the mortality experience among the Ranch Hand cohort was last assessed through 1999, and due to the age of the cohort, an updated mortality assessment would greatly improve statistical power. As such, the objectives of this study are to (1) obtain the serum TCDD data collected over time and updated mortality data for the Ranch Hand and Comparison participants; (2) use PBPK models to refine the exposure assessment; (3) conduct exposure–response modeling to assess potential mortality risk from TCDD exposures. In addition to traditional models, other approaches in exposure–response modeling will be explored such as smoothing spline regression models with the use of *drsmooth* statistical package in R.

Principal Investigator: George Knafl, University of North Carolina at Chapel Hill

Title: *Effects of Dioxin Exposure for Male Air Force Vietnam Veterans on Reproductive Outcomes*

Assets Used: Electronic data only

Abstract: The Air Force Health Study investigated the impact of herbicide exposure, especially to dioxin, on the health, survival, and reproductive outcomes of male Air Force veterans of the Vietnam War that were matched with a comparison group of Vietnam Air Force veterans not involved in those activities. Baseline data were collected in 1982. Medical records, including birth certificates, newborn clinic records, health records, and death certificates, were reviewed for all children to determine the occurrence of adverse reproductive outcomes, as well as reported adverse reproductive outcomes by parents. Serum dioxin levels were used as a measure of herbicide exposure. The effect of dioxin was based primarily on the classification of participants into Ranch Hands and comparison groups. It was concluded that the data collected prior to January 1990 provided little or no support for an adverse association between paternal dioxin exposure and reproductive outcomes. However, more complete data collected up to 2002 are available. More complete analyses of all available data have the potential for identifying adverse dioxin exposure effects on reproductive outcomes not identified in earlier AFHS analyses. Consequently, we propose to conduct more complete analyses of the effect of dioxin exposure on reproductive outcomes for all study participants combined, not differentiated by exposure status, based on all available data, considering a wide variety of potential covariates including those accounting for missing data, and allowing for the possibility of a threshold for an adverse effect on reproductive outcomes with models determined adaptively using an objective model selection criterion.

Principal Investigator: Jack Mandel, Exponent

Title: *The Reanalysis of the Ranch Hand Data*

Assets Used: Electronic data only

Abstract: This study has three aims. First, examine the association between serum TCDD level and specific health outcomes in the AFHS population using a different set of assumptions and a different statistical approach than previously applied to these data, specifically testing whether the use of "as-measured" (not log transformed) serum TCDD levels and less restrictive dose–response models would provide a more precise, reliable and valid estimate of risk in the AFHS population. The second aim is to evaluate the impact of using sub-groups of the AFHS population and various comparison groups that differ from the comparisons defined in the original study protocol on the reliability and validity of the reported risks for cancer and diabetes associated with elevated serum TCDD levels. The third aim is to determine whether the rates of change in repeated laboratory measures and clinical indicators assessed in each examination cycle for an individual study participant are associated with the rate of change in the as-measured serum TCDD level.

Principal Investigator: Allan Mazur, Syracuse University

Completed

Assets Used: Electronic data only

Mazur, A., R. Westerman, A. Werdecker, and U. Mueller. 2014. Testosterone and Type 2 Diabetes in Men. *Aging Male* 17(1):18-24.[1]

Abstract: Objective: To assess from observational data if low testosterone in men is an independent risk factor for high fasting glucose (FG) and for a diagnosis of type 2 diabetes (T2D). Methods: Multivariate analysis of data from 991 male U.S. Air Force veterans who completed six medical examinations over 20 years. Results: Low testosterone was moderately related to high FG, independent of age and obe-

[1]Mazur, A., R. Westerman, A. Werdecker, and U. Mueller. *Testosterone and type 2 diabetes in men*, copyright © 2014, Informa Healthcare.

sity. Low testosterone is a very weak predictor of a diagnosis of T2D. Conclusions: In men, low testosterone is an independent risk factor for high FG, comparable to aging and obesity. Low testosterone is a weak predictor of a diagnosis of T2D.

Mazur, A., R. Westerman, and U. Mueller. 2013. Is Rising Obesity Causing a Secular (Age-Independent) Decline in Testosterone Among American Men? *PLoS ONE* 8(10):e76178. doi:10.1371/journal.pone.0076178.

Abstract: The testosterone of men in industrial societies peaks in their twenties and tends to decline with increasing age. Apart from this individual-level decline, there have been reports of a secular (age-independent population-level) decline in testosterone among American and Scandinavian men during the past few decades, possibly an indication of declining male reproductive health. It has been suggested that both declines in testosterone (individual-level and population-level) are due to increasing male obesity because men in industrial society tend to add body fat as they age, and overall rates of obesity are increasing. Using an unusually large and lengthy longitudinal dataset (991 U.S. Air Force veterans examined in six cycles over 20 years), we investigate the relationship of obesity to individual and population-level declines in testosterone. Over 20 years of study, longitudinal decline in mean testosterone was at least twice what would be expected from cross-sectional estimates of the aging decline. Men who put on weight intensified their testosterone decline, some greatly so, but even among those who held their weight constant or lost weight during the study, mean testosterone declined 117 ng/dl (19%) over 20 years. We have not identified the reason for secular decline in testosterone, but we exclude increasing obesity as a sufficient or primary explanation, and we deny the supposition that men who avoid excessive weight will maintain their youthful levels of testosterone.

Principal Investigator: Adam Mitchell, Emory University and Atlanta VA Medical Center, Department of Veterans Affairs

Title: *Identifying Novel Biomarkers of Vulnerable Coronary Artery Disease: The Air Force Health Study*

Assets Used: Electronic data and biospecimens

Abstract: Identification of patients at high risk for primary cardiovascular events (myocardial infarction and coronary death) and innovative prevention strategies could reduce the incidence of ischemic heart disease, a leading cause of morbidity and mortality. This project aims to identify novel epigenetic and metabolic biomarkers associated with future cardiovascular events. MicroRNAs are short, noncoding RNAs that function as posttranscriptional gene regulators through targeted binding of messenger RNAs. Because microRNAs circulate in the blood and are altered in disease states, they may serve as useful biomarkers. The serum metabolome also shifts with disease, and other groups have found metabolites associated with adverse cardiovascular outcomes. We hypothesize that changes in circulating microRNAs and metabolomics profiles accompany, and likely contribute to, the development of vulnerable atherosclerotic disease. Using a nested case-control study design within the AFHS, we will characterize circulating microRNAs and metabolomics signatures associated with impending cardiovascular events. Study subjects identified as having medical record verified myocardial infarction or coronary death will serve as cases, and serum samples will be obtained from the clinical visit that immediately preceded their coronary event. Control subjects will be selected using incidence density sampling whereby a control subject with a calculated cardiovascular risk similar to each case will be sampled at the same time point as the matched case. Comprehensive microRNAs profiling will be performed using deep sequencing on 50 cases and 50 controls to discover microRNAs associated with cardiovascular events. The top 20 differentially expressed microRNAs will be validated in the remainder of the cohort with qRT–PCR. High-Performance Metabolic Profiling capable of measuring > 20,000 metabolites will be performed on all samples. The effect of serum microRNAs and metabolite levels on the occurrence of cardiovascular events will be compared within case–control matched sets using conditional logistic regression analysis. Moreover, development of an aggregate score by combining multiple microRNAs and metabolites that potentially improves the prediction of cardiovascular events will be explored by discrimination analysis. At the conclusion of these studies, we will have comprehensively characterized the circulating microRNAs transcriptome and metabolome in humans vulnerable to primary cardiovascular events. We believe these data will allow us to identify a signature of cardiovascular risk that could enhance current cardiovascular disease risk prediction models. In addition, identification of microRNAs and metabolic derangements could uncover new signaling pathways that are important in the progression of atherosclerosis proximal to the development of adverse clinical outcomes.

Principal Investigator: Kenneth Ramos, University of Louisville

Title: *Detection of L1 Protein in Ranch Hand Biospecimens*

Assets Used: Electronic data and biospecimens; pilot funding awarded

Abstract: The Air Force Health Study is one of the most exhaustive epidemiologic investigations of adverse human health effects in Vietnam veterans possibly related to herbicide exposure during aerial application of herbicides. While several herbicides were used for crop destruction, most studies have focused on Agent Orange. Questions related to the health effects of chemicals present in Agent Orange and the mechanisms involved in pathogenesis remain unanswered. Of particular interest are findings linking Agent Orange exposure to various types of cancer. As such, the study of cancer biomarkers would be highly beneficial given that veterans exposed to Agent Orange may be at heightened risk for malignancy. In this application, studies are proposed to measure L1 ORF1p (Open Reading Frame 1 protein encoded by the Long Interspersed Nuclear Element-1) as a serum biomarker of genomic instability and cancer outcomes in veterans exposed to Agent Orange. An ELISA (enzyme-linked immunosorbent assay) platform will be used to measure the abundance of L1 ORF1p in serum specimens of veterans who participated in the AFHS. ORF1p is a 40 kDa RNA binding protein that is essential for L1 retro-element mobilization and pathogenesis. Our recent studies have shown that levels of L1 ORF1p are elevated in plasma samples of cancer patients and therefore, we are interested in further evaluating the utility of this biomarker in cancer evaluation. L1 reactivation causes genetic instability secondary to DNA strand breaks, insertion mutations, altered splicing, increased frequency of recombination, and loss of transcriptional control of genes in the vicinity of insertions. One or more of these alterations has been associated with cancer. We have shown that environmental hydrocarbons reactivate L1 in somatic cells via epigenetic mechanisms that involve removal of retinoblastoma-associated repressor complexes and recruitment of proteins that mediate transcriptional activation of L1. Epigenetic reactivation of L1 reprograms the genome to induce oncogenic phenotypes and/or facilitate insertion mutations that disrupt genome architecture and function. Thus, measurement of ORF1p may be a sensitive biomarker of L1 activity and a proxy of genomic instability and cancer predisposition in human populations. The appearance of ORF1p in human serum reflects protein leakage into the extracellular compartment after apoptotic or necrotic cell death following toxic injury or disease. We hypothesize that human exposure to Agent Orange is associated with increased levels of L1-encoded ORF1p in serum, which in turn, correlates with increased incidence of cancer outcomes in Ranch Hand veterans. This hypothesis will be tested in two specific aims designed to measure L1 ORF1p in serum samples from cycles 1 and 4 of veterans exposed to Agent Orange and correlate serum levels of L1 ORF1p with TCDD levels, history of malignancy, and clinical outcomes. Measurement of L1ORF1p in serum may identify participants who have been diagnosed with cancer or those who are at increased risk of cancer development or malignant progression.

Principal Investigator: Web Ross, Pacific Health Research and Education Institute, Department of Veterans Affairs

Title: *Parkinson's Disease and Premotor Features of Parkinson's Disease in the Air Force Health Study*

Assets Used: Electronic data only; pilot funding awarded

Abstract: Evidence from epidemiologic and basic science research suggests pesticides contribute to Parkinson's disease risk. Specifically, 2,4-dichlorophenoxyacetic acid (2,4-D) and 2,4,5-trichlorophenoxyacetic acid (2,4,5-T), the two constituents of Agent Orange, have been implicated in two separate studies, prompting the Institute of Medicine to acknowledge a suggested association of pesticides with

Parkinson's disease. Additionally, the Department of Veterans Affairs (VA) has added Parkinson's disease to the list of illnesses possibly caused by Agent Orange making Parkinson's disease a service-connected condition for Vietnam veterans. There have been no studies of Agent Orange exposure and Parkinson's disease in Vietnam veterans. The AFHS data provide a unique opportunity to address the association of Agent Orange exposure in Vietnam with Parkinson's disease as well as with premotor features of Parkinson's disease. The four aims seek to address key questions related to Agent Orange exposure and Parkinson's disease risk using data from the AFHS. (1) determine prevalence of Parkinson's disease in exposed Ranch Hand participants versus nonexposed comparison participants; (2) examine the association of serum dioxin with Parkinson's disease in AFHS participants; (3) determine prevalence of premotor Parkinson's disease features in Ranch Hand participants versus comparison participants; and (4) examine the association of serum dioxin with premotor Parkinson's disease features in AFHS participants. To accomplish this, the researchers will determine the prevalence of Parkinson's disease in Ranch Hands and comparison participants from self-report at exam 6 and, after 2002, from VA and Medicare datasets, and use data from exam cycles 1–6 to determine the prevalence and association of serum dioxin with premotor Parkinson's disease features in the cohort.

Principal Investigator: David Roth, San Diego Healthcare System, Department of Veterans Affairs

Title: *Caveolin's Role During Healthy Aging*

Assets Used: Electronic data and biospecimens; pilot funding awarded

Abstract: Heart failure is the most common cause of death in older patients and is often caused by cardiovascular disease. Cholesterol is an important component of the cell membrane and is an essential element of membrane lipid rafts. Caveolae, a subset of membrane lipid rafts, serve as signaling centers and contain caveolin proteins, cholesterol, and sphingolipids. This study aims to determine if there is an association between caveolin protein expression levels in serum and adipose tissue and the development of heart failure. The role of caveolin and membrane lipid raft expression and serum cholesterol levels on the development of heart failure will be tested by investigating serum and adipose tissue samples from a pilot cohort of nonherbicide exposed participants: 15 participants that developed heart failure and 15 matched control participants who did not develop heart failure. Investigation will also include the role of serum cholesterol and tissue cholesterol levels on the development of heart failure by testing the total cholesterol content in whole serum and in a subset of whole adipose tissue samples. Correlation analysis between caveolin protein expressions, cholesterol levels, and the incidence of heart failure will be performed.

Principal Investigator: David Seldin, Amyloid Treatment & Research Program, Boston University School of Medicine and Boston Medical Center

Title: *Incidence of Abnormal Free Light Chains and Other Markers of Light Chain Amyloidosis in Veterans Exposed to Agent Orange: A Pilot Study*

Assets Used: Electronic data and biospecimens; pilot funding awarded

Abstract: The researchers propose detection of preclinical markers of amyloidosis in the serum of veterans exposed to Agent Orange can be used to facilitate early diagnosis and treatment. Participants that have both serum samples and adipose tissue specimens from the cycle 5 exam will be used to analyze immunoglobulin free light chains and other serum biomarkers of clonal plasma cell disease in participants with elevated free light chains. These tests will be used to attempt to determine if these participants have developed clinical features of amyloid disease by measuring known biomarkers of amyloid cardiomyopathy in the serum confirmed with adipose tissue as well as linking to mortality cause information suggests they may have died due to amyloidosis.

Principal Investigator: Youn Shim, Centers for Disease Control and Prevention

Title: *Monoclonal Gammopathy of Undetermined Significance and MicroRNAs in Ranch Hand Veterans*

Assets Used: Electronic data and biospecimens; pilot funding awarded

Abstract: Monoclonal gammopathy of undetermined significance (MGUS) is a premalignant disorder resulting from unregulated clonal expansion of antibody-secreting B-cells. Individuals with MGUS can be identified by detecting monoclonal immunoglobulins (M-protein) and unbalanced free light chains in serum. The prevalence of MGUS increases with age and is about 4.8% in men aged 50 years and older. MGUS confers significantly increased risk for multiple myeloma, progressing at a rate of 1% per year. Multiple myeloma is considered to have suggestive evidence of an association with exposure to herbicides, including Agent Orange and its contaminant TCDD, among Vietnam veterans. TCDD is known to affect microRNA levels by binding of the miR-191 promoter to the AhR transcription factor. TCDD may alter microRNA levels that are involved in regulating apoptosis and other critical processes of cell proliferation. Malignant cells have been found to selectively package their microRNAs into exosomes for secretion, and altered levels of certain microRNAs are reported in multiple myeloma as well as in MGUS.

MicroRNAs, free light chains, and M-protein have long-term stability in serum stored under a wide range of conditions. Our study will be the first to investigate MGUS and microRNAs using sensitive state-of-the-art laboratory methods to analyze stored serum specimens of Air Force Health Study (AFHS) participants. The purpose of this study is to determine whether exposure to Agent

Orange and TCDD increased the risk for MGUS in Ranch Hand veterans, by using serum samples stored by the AFHS during the 2002 investigation. We will also explore whether TCDD influences the progression of MGUS by testing for three novel biomarkers: M-protein concentration, free light chain ratio, and a microRNA panel. For this purpose, we will use serum specimens stored at three time points (1982, 1992, and 2002).